Integrated Micro-Ring Photonics

Integrated Micro-Ring Photonics

Principles and Applications as Slow Light Devices, Soliton Generation and Optical Transmission

Iraj Sadegh Amiri
Photonics Research Centre, University of Malaya, Kuala Lumpur, Malaysia

Abdolkarim Afroozeh
The Department General of Fars Province Education, Iran
Young Researchers and Elite Club, Jahrom Branch, Islamic Azad University, Jahrom, Iran

Harith Ahmad
Photonics Research Centre, University of Malaya, Kuala Lumpur, Malaysia

CRC Press
Taylor & Francis Group
Boca Raton London New York

CRC Press is an imprint of the
Taylor & Francis Group, an **informa** business

A BALKEMA BOOK

CRC Press
Taylor & Francis Group
6000 Broken Sound Parkway NW, Suite 300
Boca Raton, FL 33487-2742

First issued in paperback 2019

© 2017 by Taylor & Francis Group, LLC
CRC Press is an imprint of Taylor & Francis Group, an Informa business

Typeset by MPS Limited, Chennai, India

No claim to original U.S. Government works

ISBN-13: 978-1-138-02783-1 (hbk)
ISBN-13: 978-0-367-87370-7 (pbk)

Library of Congress Cataloging-in-Publication Data

**Visit the Taylor & Francis Web site at
http://www.taylorandfrancis.com**

**and the CRC Press Web site at
http://www.crcpress.com**

Table of contents

Preface

In this book, interferometric system of Micro-Ring Resonator (MRR) for chaotic signal generation using the fiber optic is presented. The introduction and theory of the soliton generation and propagation in fiber optics is presented in chapters one and two. This MRR system can be incorporated with an optical add/drop filter system, which constitutes an efficient system to generate the soliton comb, where it has several applications in optical communication. The filtering process of the chaotic signals occurs during the round-trip of the pulse within the ring resonators. The theory and analysis of the single MRR, add/drop MRR and cascaded MRRs is presented in chapter three. The physics and fabrication of the MRR systems are presented in the chapter four. The MRRs used as optical modulators is presented in chapter five, and the application of MRR system in optical communication is presented in chapter sixth. Chapter seven presents the techniques of generating slow and fast pulse using micro-ring resonator. Here different semiconductor materials are examined. The slow light is generated in micro-ring resonator systems with bright soliton and Gaussian pulse as an input. The generation of slow light is analyzed mathematically through the scattering matrix method and simulated by the MATLAB program. This approach allows for the aspects of the generated output, that being high-speed and multi-channel modulation using cascaded silicon micro-ring modulators, to be clearly shown. Also, this book contains comprehensive details of the optoelectronic components that comprise the feedback system, and their integration onto a single device using CMOS-compatible processes and materials. The slow light pulses generated in linear and nonlinear micro-ring resonator configurations are examined in terms of the effect of physical parameters such as ring radius, numbers of ring, core area and coupler coefficients on the output pulse. Using the series of MRR connected to the add/drop MRR system, dark and bright soliton pulses possessing Full Width at Half Maximum (FWHM) and Free Spectral Range (FSR) of 10 pm and 163 pm respectively are generated, where generation of 83 fs is obtained. The soliton pulse generation and bandwidth manipulation of the pulse is performed and analyzed using the MRR system. With the Gaussian laser input, the multi-solitons with FWHM = 50 pm and FSR = 1440 pm can be generated. Here, analysis of the FWHM is performed respect to variation of the MRR's radius and coupling coefficient. As application in optical communication systems, soliton with FSR = 6.66 MHz can be generated and used as optical carriers to be modulated and transmitted via a wired/wireless system. Using the PANDA MRR system, the input Gaussian laser pulses with power of 0.45 W are inserted into the system. The central wavelength of the input power has been selected to $\lambda_0 = 1.55\,\mu\text{m}$

where the nonlinear refractive index of the medium is $n_2 = 1.3 \times 10^{-17}\, m^2\, W^{-1}$. The chaotic signals with bandwidth of 24 pm can be transmitted through the fiber optic up to 195 km. As result, signals with 600 fm bandwidth could be trapped within the system. The nano bandwidth tweezers can be generated using a system known as Half-Panda and transmitted within an optical transmission link. These signals can be connected to the fiber optic with a length of 100 km, where transmission of tweezers can be performed. Here, signals with FWHM of 33 nm and FSR of 50 nm are obtained and transmitted. Using these signals, the quantum photon via an optical communication link can be performed. The input Gaussian soliton is used to control the output signals. The receiver will detect the entangled photon pair signals and transmit them via wired/wireless as quantum bits. The work also represent smallest optical tweezers signals with FWHM of 4.2 nm is obtained where the FSR of 50 nm is obtained.

Acknowledgements

Iraj Sadegh Amiri would like to acknowledge the University of Malaya for providing funding for this research under the grants LRGS (2015) NGOD/UM/KPT, UMRG (RP029B-15AFR), GA010-2014 (ULUNG) and RU007/2015.

List of figures

List of tables

Chapter 1

Soliton signals propagating in fiber waveguides and slow light generation

1.1 FIBER WAVEGUIDES

Optical beams have a natural tendency to become diffracted while propagating in a uniform medium. The beams diffraction can be compensated by beam refraction when the refractive index is increased. The use of optical waveguides is a key way to present a balance between diffraction and refraction if the medium is uniform regarding to the direction of propagation. Therefore the outcome propagation of the light is controlled in the transverse direction of the waveguide, and it is described by the concept of spatially localization of the electric field in the waveguide (Snyder & Love, 1983).

The first waveguide was proposed by J. J. Thomson in 1893, and experimentally verified by O. J. Lodge in 1894. Analysis of the propagating modes was executed mathematically by Lord Rayleigh in 1897 within a hollow metal cylinder. In April 1957, scientists tried to achieve maser-like amplification of visible light. In November of 1957, Gordon Gould, an American physicist (credited with the invention of the laser) could make an appropriate optical resonator by using two mirrors in the form of a Fabry-Perot interferometer. Unlike other designs, this new design would produce a narrow, coherent, intense beam. The gain medium could easily be optically pumped in order to achieve a necessary population inversion. He also considered pumping of the medium by atomic-level collisions, and expected many potential applications of such a device (Hammond *et al.*, 2002). Lasers have a wide range of applications, such as lidar, ladar, and communications.

Typically, an optical fiber consists of a transparent core surrounded by a transparent cladding material with a lower refractive index. Light is confined in the core by total internal reflection so it acts as a waveguide. Fibers which can be used for many propagation paths or transverse modes are called Multi-Mode Fibers (MMF) whereas those which can only support a single mode are called Single Mode Fiber (SMF). The MMF types of fiber generally have a larger core diameter and can be used for short distance communication systems while SMF fibers are used for long distance communication, which in this case is considered in excess of 1,050 meters (3,440 ft) (Hecht, 2010).

1.2 OPTICAL SOLITON

Optical solitons are localized as electromagnetic waves that propagate in nonlinear media resulting from a balance between nonlinearity and linear broadening due to dispersion and/or diffraction (Kouhnavard *et al.*, 2010b; Shojaei & Amiri, 2011b).

There are five types of nonlinear medium, as categorized by their behaviour in accordance with the Kerr law, power law, parabolic law, dual-power law and the log law. In the presence of dispersive perturbation terms, the phenomena of optical soliton cooling are also observed. Initially, the term soliton referred to the particle-like nature of solitary waves that remain intact even after common collisions (Abdullaev & Garnier, 2005). The first observation of soliton was done by Scott Russel on the Edinburgh-Glasgow canal in 1834. He observed that a wave travelling through a canal without lost and major changes of its shape (Chiao *et al.*, 1964). The first experimental observation of soliton (Mollenauer *et al.*, 1980) occurred via microscope after a soliton was generated when the mode-locked color center laser's output was coupled into the fiber, and the fiber's output into an autocorrelator.

This observation appears to disagree with the nonlinear theory of Airy published in 1845, which predicted that a wave of finite amplitude cannot transmit without a change of its shape. According to his theory the wave should attenuate. The problem was solved by Joseph Boussinesq (Narahara & Nakagawa, 2010) in 1871. In 1876 Lord Rayleigh (Sander & Hutter, 1991) independently was able to show that in a solitary wave the increase in local wave velocity associated by finite amplitude is balanced by the decrease associated with dispersion. In 1895, Korteweg de Vries (Israwi, 2010) developed a model which can explain the unidirectional propagation of the waves of long wavelength in water with relatively shallow depth. This equation now is known as KdV. However the properties of solitons are not clearly understood until several mathematical models were introduced. The inverse scattering method was developed in the 1960s and it was able to explain the properties of soliton. The mathematical solution of soliton as KdV was found by Zabusky and Kruskal in 1964 (El *et al.*, 2009).

In 1973 it was discovered that optical fibers can support dark solitons when the Group-Velocity Dispersion (GVD) is "normal". Hasegawa and Tappert could solve and explain the Non-Linear Shrödinger (NLS) equation and the theory of optical soliton (Zhang *et al.*, 2011). The first generation of spatial solitons was reported in 1974 by Ashkin and Bjorkholm (Wise, 2001) in a cell filled with sodium vapor. Only a decade later, Mollenauer performed the first experiment of soliton propagation in optical fibers, since low loss fibers could not be suitably adapted for use at that time (Mollenauer & Smith, 1988). Temporal dark solitons became very interesting during the 1980s (Stratmann & Mitschke, 2005). During the 1990s, many other kinds of optical solitons such as spatiotemporal, Bragg, vortex, vector and quadratic solitons were discovered.

In the most recent overview of experimental observations of spatial optical solitons, some materials display large optical non-linearities when their properties are customized by the light propagation (Fischer *et al.*, 2006). This is particularly the case for non-linearity characteristics, when a resulting change in the refractive index of the medium means the beam can become self-trapped and propagates unchanged exclusive of any external waveguiding structure. These types of stationary self-guided beams are known as spatial optical solitons (Stegeman & Segev, 1999).

In the 1970s, Hasegawa began to realize that the NLS equation was appropriate for the calculation of pulse propagation in optical fibers, and that they should therefore support solitons. In a seminal work published in 1973 (Hasegawa & Tappert, 1973a), he and co-author Frederick Tappert showed how the NLS equation applied to single-mode fibers, derived the essential properties of the corresponding solitons.

In supporting numerical simulation they showed that the solitons were stable and robust. It is noteworthy that at the time, fibers having low loss in the region of anomalous dispersion ($\lambda > 1300$ nm) did not exist. Hasagewa and Tappert followed up almost immediately with another paper (Hasegawa & Tappert, 1973b) describing dark solitons, i.e., sech-shaped holes in a CW background, which could exist in the presence of normal dispersion. For a number of practical reasons, however, the dark solitons have never been used for transmission.

1.3 RING RESONATORS

The shape of fiber is adjustable, thus the ring resonators can be made and used to resonate selective wavelength or can be used as filters. MRRs can be shaped from nanoscale photonic waveguides (Afroozeh *et al.*, 2011b). Ring resonators are employed to generate signals used for optical communication applications, where they can be integrated in a single system (Ali *et al.*, 2010g; Amiri *et al.*, 2012j). Optical MRRs recently are interesting subject in the area of integrated optics because of their unique aspects such as compactness, low cost, tenability, and easy integration on a chip with other photonic devices, having a variety of applications such as optical filter, optical switch, optical modulator, optical delay line, dispersion compensator, and optical sensor (Little *et al.*, 1997b).

Since ring resonators are used to support the travelling of wave resonant modes, a single ring may be applied to completely extract a particular wavelength from a signal bus, they are therefore ideal candidates for very large-scale integrated (VLSI) photonic circuits, since they provide a wide range of optical signal processing functions while being ultra-compact (Daldosso & Pavesi, 2009). MRRs possess clear advantages over other candidates when they are used as a filter system (Liang *et al.*, 2011). The type of semiconductor microring is used widely to enhance the nonlinear optical effects which are proposed and investigated (Absil *et al.*, 2000).

Ring resonators are not used only in optical networks, but they have recently been presented to be used as sensors, filters and biosensors (Amiri & Ali, 2013d; Amiri *et al.*, 2014b). There are many research works on the fabrication and characterization of integrated ring resonators in a variety of material systems. Rafizadeh (1997), wrote the first thesis on integrated ring and disk resonator filters in which fabrication and characterization of devices with diameters smaller than $10\,\mu$m is investigated in the material system AlGaAs/GaAs. First theoretical works were written by Hagness (1998), where the Finite Difference Time Domain (FDTD) analysis of the ring and disk resonators was presented. In 2000 Absil wrote a thesis based, on the material system AlGaAs/GaAs, where a vertical and lateral multiple coupled ring resonator configuration could be fabricated and characterized. Most of the research works on the ring resonator in the micro and nano size scale have been done since 2000. The security and high capacity of optical communication network is the major concern in the field of nano photonics.

1.4 APPLICATIONS OF RING RESONATOR SYSTEMS

Micro and nanoscale devices have been used widely in information technology such as telephone handsets (Amiri *et al.*, 2014c). One of the important parts of this device is an antenna. Using a small antenna with good performance is necessary, where to date, nano-communication has become an interesting field in many applications such as communication and networks. A novel nano-communication system design is presented by Amiri *et al.* in which photonic spins in a PANDA ring resonator are employed (Amiri *et al.*, 2015a). These spins are generated using soliton pulse within a PANDA system. The magnetic field is introduced by using an aluminum plate coupling to the MRR, in which the spin-up and spin-down states are induced, where finally, the photonic dipoles are formed. The advantage of the proposed system is that powerful simple and compact nano-antenna can be fabricated. In addition, optical dipole can be used for further research such as dynamic dipole, dynamic torque, nano-motor, spin communicated and spin cryptography, etc. (Thammawongsa *et al.*, 2012).

The use of data and information in optical communication is growing day by day. Therefore, the security of data is a major concern, where there are a lot of techniques which can be used to protect the secret data or information. Up to date, a quantum technique is recommended to provide such a requirement (Ali *et al.*, 2010a; Amiri *et al.*, 2012e). A new concept of quantum cryptography using dark-bright soliton conversion behaviors within a nonlinear ring resonator (PANDA ring resonator) is presented by Amiri *et al.* In this research orthogonal soliton is established among the soliton conversion (Amiri *et al.*, 2012o). The advantage of this research is that long distance quantum communication and high capacity quantum communication can be performed using the powerful entangle soliton. In application, the high capacity quantum communication is variable by using the multi variable entangled solitons (Tunsiri *et al.*, 2012).

Chaotic signals have some properties such as broadband, orthogonality and complexity aspects, which prompt researches in the areas of nonlinear science, communication technology and signal processing. The concern in chaotic communications was due to the foreseen good properties of the chaotic signals in the fields of security systems or broadband multiple access systems (Alavi *et al.*, 2013a; Amiri *et al.*, 2014j). The possibility of employing chaotic signals to carry information was first studied in 1993. Encoding is the process of adding the correct transitions to the message signal in relation to the data that is to be sent over the communication system. Fiber optic sensors and micro structured fibers hold great promise for integration of multiple sensing channels. Nonlinear behavior of light inside a MRR takes place when a strong pulse of light is inserted into the ring system. Chaotic controls have been used in a great number of optical, engineering and biological designed systems.

Optical communication has become an interesting area in photonics over the last two decades. It is very attractive especially when it uses quantum cryptography in a network system as reported by Amiri *et al.* Quantum keys can be constructed as an aspect of the required transmission information and so provide the perfect communication security. Amiri *et al.* showed that quantum security could be performed via the optical-wired and wireless link. Several research works have proposed various techniques of quantum cryptography, for which the systems of MRR are still complicated (Amiri *et al.*, 2014a). Amiri *et al.* proposed a new quantum key distribution rule in which carrier information is encoded on continuous variables of a single photon. In this mechanism, a person wishing to transmit a message securely can randomly encode information on either the central frequency of a narrow band single-photon pulse or the time delay of a broadband single-photon pulse. Liu and Goan studied the entanglement evolution under the influence of non-Markovian thermal environments

(Liu & Goan, 2007). The continuous variable systems could be two modes of electromagnetic fields or two nano-mechanical oscillators in the quantum domain, whereby there is no process that could be performed within a single system.

To generate a spectrum of light over a broad range, an optical soliton pulse is recommended as a powerful laser pulse that can be used to generate chaotic filter characteristics when propagating within MRRs. Therefore, the capacity of the transmission data can be secured and increased when the chaotic packet switching is employed (Jalil et al., 2011). In this book, we obtain localized spatial and temporal soliton pulses to form the transmission characteristics of the soliton signals. The MRR system is used to trap optical solitons in order to generate the entangled photon pairs that are required for quantum keys. Here, generation of the localized ultra-short soliton pulses for continuous variable application is demonstrated. The system of quantum key generation can be implemented within the wireless networks. Thus, the links can be set up using the optical soliton, generated by the technique called chaotic filtering scheme in which required signals can be selected and used. The device parameters are simulated according to the practical device parameters, where the results obtained have shown that the entangled photon pairs can be utilized within the MRR device.

A network system can be designed to provide transmission of secret data with the highly efficient transmission of soliton signals based on the OFDM application. In this work, the optical soliton in a nonlinear fibre MRR system is analysed in order to generate a high frequency band of pulses to be multiplexed with generated logic codes from chaotic signals, using an OFDM technique (Amiri et al., 2014c). Control of the process can be achieved by controlling the parameters of the system, such as round trip, input power, coupling loss, coupling coefficient, the ring's radius, and linear/nonlinear refractive indices.

Dark-Gaussian soliton controls within a semiconductor add/drop multiplexer has numerous applications in optical communication. Nano optical tweezers technique has become a powerful tool for manipulation of micrometer-sized particles/photons in three spatial dimensions. It has the unique ability to trap and manipulate molecules/photons at mesoscopic scales with widespread applications in biology and physical sciences. The output is achieved when the high optical field is set up as an optical tweezers. For communication's application purposes, the optical tweezers can be used to generate entangled photon within the proposed network system (Amiri et al., 2012l). The tweezers in the forms of valleys or potential wells are kept in the stable form within the add/drop filter. MRR's are a type of Fabry-Perot resonators which can be readily integrated in array geometries to implement many useful functions. Several emerging technologies, such as integrated all optical signal processing and all-optical quantum information processing, require interactions between two distinct optical signals.

Dark-bright soliton controls within a semiconductor add-drop multiplexer allow for numerous applications in optical communication (Amiri, 2014). MRRs can be readily incorporated into an interferometer system to produce a specific intensity output function. In this book we present the concepts and techniques on chaotic signal trapping within a fiber optic system for optical devices used in optical communication. It demonstrates optically trapping of microparticles on silicon MRRs. The promising techniques of light trapping and transportation within the fiber optic have been reported in both theory and experimental demonstrations respectively

(Amiri & Ali, 2013). A PANDA ring resonator can be used to generate chaotic signals. By using the proposed system, the transceiver can be integrated and performed by using a single device. These resonators hold great promise for use as optical switching systems.

Exciting new technological progress, particularly in the field of tunable narrow band laser systems, optical trapping and storing and the MRR interferometers, provide the foundation for the development of new transmission techniques. Here, the highly chaotic signals can be generated and sliced into ultra-short single and multi-soliton pulses. The trapping of optical soliton pulses in pico and femtometer can be performed using the proposed system. In applications, the stored ultra-short optical signal can be used to generate optical quantum memory, where the multi-soliton generation is the advantage for the systems of ring resonators (Amiri & Ali, 2014a). Beside improvements in efficiency and beam quality these soliton sources provide short and ultra-short pulses, leading to improved process efficiencies and new fields of laser application. The soliton pulses are so stable that its shape and velocity is preserved while travelling along the medium. The increase in communication capacity is obtained by using more available channels and large bandwidth.

Additional information regarding these kinds of behaviors in a MRR evidently are defined by Amiri *et al*. Nonetheless, aside from the penalties of the nonlinear behaviors of light traveling within the fiber ring resonator, there are several benefits that can be employed by the communication methods in order to examine the obtained result. The chaotic behavior which has been employed to make the benefit within digital or optical communications (Amiri *et al*., 2012c). The ability of chaotic carriers to synchronize in a communication system is valid. Recently, Amiri *et al*. have reported the successful experimental research based on generating and transmission of chaotic signals using an optical fiber communication link. We propose a system for chaotic signal generation and cancellation using a MRR fiber optic system, where the required signals of single bandwidth soliton pulse are recovered and manipulated using an add/drop system. Results show particular possibilities with this application. Also, effects of coupling coefficients on the bandwidth of the single soliton pulse are investigated here.

1.5 INTRODUCTION OF SLOW LIGHT

Many scientists have taken keen interest to measure speed of light with high accuracy and precision (Gauthier & Boyd, 2007). Controlling the speed of light is important for many applications (Mork, 2008). Slow light refers to the control of the light velocity in a particular medium. The term slow light refers to the propagation of an optical signal through a medium with a speed considerably less than the speed of light in vacuum. The nature of light matter interactions is also important in order to understand light propagation in a particular medium, as it is dependent on the nature of the medium material (Born & Wolf, 1999).

Controlling the speed of a light signal has many potential applications in fiber optic communication networks, optical processing and quantum computing (Stenner *et al*., 2003). Optical signals can be modulated to slow down or increase the light speed using nonlinear micro ring resonators. The phenomenon of speed of light is still of great interest due to potential applications associated with speed of light using different media in terms of slow light.

Soliton pulse in the form of Gaussian beam and bright soliton is used to generate slow light. Controlling the group velocity of light pulses is useful to achieve various functionalities in an all optical packet switched network. In an optical network, light pulses are used to transmit digital information. The demand for higher bandwidths for future internet services requires all-optical networks with ultra-high speed photonic switching. At the same time, such an all-optical network can drastically reduce the energy requirements and therefore, the slow light is a footprint of communication and information systems (Tucker *et al.*, 2005).

The slow light effect has many important applications and is a key technology for all optical networks such as optical signal processing (Henker, 2010; Zadok *et al.*, 2011), the radio frequency-photonics (Henker *et al.*, 2008b), nonlinear optics and spectroscopy in the time (McMillan *et al.*, 2010). Slow light techniques can be used for future optical communication systems, including optical buffering, data synchronization and optical memories (Boyd *et al.*, 2006). Tunable optical buffers (TOBs) are key components in optical communication. The group velocity of light should be controlled in to order to realize the tunable optical buffers (Wang *et al.*, 2009b). One of the most important components in optical communications and signal processing is a controllable variable optical memory. Throughout the storage time in a buffer, optical data can be reserved in optical format without being converted into the electronic format. The optical buffer has the ability to store and release optical data at a rapid rate from an external command. An optical buffer can be defined as the storing of a signal for a time T with low distortion and attenuation. The delay time is variable and controllable (Zalevsky *et al.*, 2005). Slow light can be used for applications such as time resolved spectroscopy, microwave photonics and nonlinear optics (Sales *et al.*, 2010). The generation of slow light in MRRs is based on the nonlinear optical fibers (Biswas & Pati, 2011). To decrease the speed of light, a far shorter device is required. If the slow light effect can be switched off while the pulses are in the device, then the retrieval time can be improved.

1.6 SLOW LIGHT

Recent research has established that the velocity of light pulses can be controlled through micro ring resonators. Extremely slow propagation can be achieved through different models. This book will also give a description of the underlying physical processes leading to the modification of the velocity of light by micro ring resonators.

To understand the concept of slow light, it is important to differentiate between the phase velocity and the group velocity of a light field. The group velocity is the velocity with which a pulse of light propagates through a material system. The slow light depends on the value of the group velocity (v_g) in comparison to the velocity of light c in a vacuum. Slow light refers to the situation $v_g < c$ (Garrett & McCumber, 1970).

Different methods and systems have been proposed to generate slow light via semiconductors, optical amplifiers and fibers (Thévenaz, 2008). In principle, all these method are based on three major categories which relies on the micro ring resonators as a waveguide, phase shift and the time frequency coherence of signals.

In this study the first and second categories are investigated. The velocity of the pulses can be changed by creating artificial tailored resonances or dispersions inside the material. Hence, there is a continuous alteration of the group velocity of light pulses. Therefore, the time delay in the material with constant length can be controlled externally within a particular range.

1.7 BACKGROUND OF SLOW LIGHT GENERATION

The phenomenon of slow light has received considerable attention since the spectacular experiment has been performed. The interest to control the speed of light has been motivated by the on-going research on the fundamental nature of light matter interactions as well as the possibilities to use these properties for various applications. In particular, the use of slow light effect for optical buffering. There are basic physical limitations due to bandwidth and the delay time of the pulse. In contrast, the slow light is used to control the phase of microwave signals as intensity modulation of an optical signal which have many applications within the microwave photonics (Capmany *et al.*, 2005).

In 2008, Pornsuwancharoen *et al.* proposed a simple system consists of three rings that can be used to generate slow light and also used to stop and store the light (Pornsuwancharoen *et al.*, 2010). In 2009, Suhailin *et al.* has demonstrated stopping and storing light pulses using a system consists of an erbium-doped fiber amplifier (EDFA), a semiconductor optical amplifier (SOA) and a fiber ring resonator. In 2010, Pornsuwancharoen *et al.* generated fast and slow lights using microring resonator (MRR) for network systems. Chaiyasoonthorn *et al.* generated fast light using two micro ring resonators and a nanoring resonator with different radii and coupling coefficients. Slow light generation in a silicon nitride-based ring resonator has been experimentally implemented by Uranus *et al.* (2007).

1.8 PROBLEM STATEMENT

Over the last few years, the study of slow and fast light is becoming an important and attractive research area. At present, the proposed systems such as electromagnetically induced transparency, optical fiber and fiber Bragg gratings for the control of light are still complicated which renders it difficult for a realistic implementation. Therefore, there is a need to search for a suitable device for the means to control light to a satisfactory degree. Thus this study will look into the possibility of light control via slow light generation in new micro ring resonator waveguides such as InGaAsP/InP, GaAlAs/GaAs and hydrogenated amorphous silicon. A number of linear and nonlinear micro ring resonator systems will be proposed to control the light velocity.

1.9 RESEARCH ACHIEVEMENTS

The main objective of this section is to generate fast light and slow light and investigate the optical pulse propagation using micro ring resonator system. The specific objectives of this research are as follows:

- To derive and formulate the slow light in micro ring resonator using scattering matrix method.

- To investigate transmission characteristics of cascaded microring resonators (MRRs)
- To analysis the fabrication process of the MRR systems and modulators
- To model and simulate slow light generation using MRR through the MATLAB coding.
- To generate and investigate the slow light:
- To generate soliton pulses and transmission within fiber link

  – Using nonlinear and linear configurations of micro ring resonators.
  – For bright soliton and Gaussian beam as inputs.
  – Into three different optical waveguides such as, InGaAsP/InP, GaAlAs/GaAs, hydrogenated amorphous silicon waveguides.
  – Generation of optical pulse with FWHM in femtosecond as a fast light.

1.10 SCOPE OF RESEARCH

Ring resonators are employed to generate the signals for use in optical communication and then can be integrated into a single system. Nonlinear behaviors of the signals inside the ring resonators shows an interesting phenomena in which the secured pulses with high capacity ranges can be obtained for long distance communication. This research study, involves both the numerical experiment and theoretical work based on MRR for secured communication. The theoretical part of this study uses modelling techniques based on the proposed MRR systems.

In order to write the programs, the developed equations from the ring resonator systems are converted to suitable programs using scattering matrix method function. They are used to obtain the results via MATLAB, Optisystem, and Comsol programming. In this simulation, iteration method and numerical analyses have been used. The basic equations refer to the relation between the electric fields inside single ring in the stationary state which can be obtained from the nonlinear propagation equations. The output power of the ring resonator can be expressed as a function of the number of ring circulations in micro ring resonator. Slow light can be obtained by controlling of the group velocity. Thus, an optical delay line can effectively function as an optical buffer and the storage is proportional to the variability of the group velocity. Control the delays in optical fibers under the new system as a ring resonator at telecommunication wavelengths has paved the way towards real applications for slow and fast.

1.11 SIGNIFICANCE OF STUDY

Micro ring resonators offer several advantages over the other methods and the group velocity can be controlled over a very wide range. The systems are very easy to implement and can be built using standard components of telecommunications (Arunvipas et al., 2011). Furthermore, the MRRs can be made in several kinds of materials which make the systems very flexible. The extreme speed in which the light moves, and the fact that photons do not tend to interact with transparent matter, is of enormous benefits of optical communication. It allows us to transmit data over long distances in optical fibers.

The speed of light is 3×10^8 m/s in a vacuum. This makes the possibility of transmitting information with almost no delay, even over ultra-long distances. There are strong motivations to dramatically slow down the light. In particular, the creation of more efficient and sophisticated optical communication networks in the future will probably require direct control over the speed of light in optical information propagates. For instance, a key element of such networks will be the all-optical router which is a device that controls the flow and timing of optical data without the conversion into electronic signals. Such all-optical routers will require dynamically controllable optical delays and the ability to perform short-term storage of light pulses.

The capability of controlling light in this manner is very recent. It could ultimately lead to more advanced forms of optical processing where the actual information that is sent across a network can be determined by the conditional processing of the optical inputs from many distant nodes in the network. Light pulse can be stopped and stored into the MRR and readily is available for Read Only Memory (ROM) application. The chaotic cancellation can be made using the fast and slow light method. Therefore fast and slow light can be achieved in the separation time. In practice, the key points of this application are the encrypted parameters, which will be implemented in the near future.

Slow light can be generated within the micro ring devices, which will be able to use with the mobile telephone. Therefore, the message can be kept in encrypted via quantum cryptography. Thus the perfect security in a mobile telephone network is plausible.

1.12   HISTORY OF SLOW LIGHT GENERATION

Nowadays, the phenomena of reducing the speed of light in nonlinear optics are of great importance. Also the fundamental physical interest in this approach has very high practical potential. The increasing demand on higher bandwidths for future internet services requires the development of secured communication networks with ultrahigh speed photonic switching and optical buffer. Optical soliton as an input and micro ring as a waveguide have been used to reduce or increase the speed of light. The soliton was first observed by John Scott Russell in 1834 (Stanton & Ostrovsky, 1998). Joseph Boussinesq made the important contributions to solitary wave by showing that if one ignores dissipation, the increase in local wave velocity associated to finite amplitude can be balanced by the decrease associated with dispersion. The early history of solitons has been marked by long eclipses. For nonlinear materials the basic equations to describe the formation and propagation of different types of optical solitons are well known. For weak nonlinearity the most conceptually simple solitons, such as spatial solitons or stationary self-guided beams and temporal solitons such as pulses in optical waveguides are described by a mathematically identical scalar equation (Shen, 1984; Agrawal, 2000). This equation is known as the Non-Linear Schrödinger (NLS) equation. In 2007, Yupapin *et al.* used soliton pulse into MRRs for several applications such as optical communication, enhance capacity, medical application, THz generation of radio over fiber (Al-Raweshidy & Komaki, 2002).

Optical communication systems first emerged via optical telegraph that Claude Chappe that was invented in the 1970s (Hecht, 2004). After passing a long period of time, guiding of light was demonstrated by Daniel Colladon in fiber optics in the 1840s (Bates, 2001). Fibers are used instead of metal wires because signals travel along them with less loss and are also immune to electromagnetic interference. In 1854, John Tyndall demonstrated that light could be conducted through a curved stream of water, proving that a light signal could be bent (Comtois, 2001). In 1880, Alexander Graham Bell invented his 'Photophone', which transmitted a voice signal on a beam of light. Bell focused sunlight with a mirror and then talked into a mechanism that vibrated the mirror (Hecht, 1985).

Optical fiber depended on the phenomena of total internal reflection, which enables confinement of light in one material that is surrounded by another material having a lower refractive index. Jun-Ichi Nishizawa, also proposed the use of optical fibers for communications in 1963 (Hecht, 2004). The first working fiber-optical data transmission system was demonstrated by Manfred Börner in 1965 (Hecht, 2004). Charles K. Kao and George A. Hockham were the first to promote the idea that the attenuation in optical fibers could be reduced below 20 dB/km, making fibers a practical communication medium (Palais, 1988). They proposed that the attenuation in fibers available at the time was caused by impurities that could be removed, rather than by fundamental physical effects such as scattering. They correctly and systematically calculated the light-loss properties of optical fiber, and pointed out the right material to use for such fibers silica glass with high purity.

Pornsuwancharoen *et al.* used micro ring resonator as a waveguide for optical communication in 2008. Micro resonators constructed in III–V semiconductors began seeing light in the early 1990s. Several groups demonstrated optically pumped micro disk lasers in both GaInAsP-InP and III–Nitrides using the whispering gallery mode. The smallest disks of circumference $\sim 15\mu$m were reported (Lun *et al.*, 1998a; Lun *et al.*, 1998b). Most of these early efforts did not incorporate bus waveguides and relied on fibers to collect light directly from the disk. The first GaAs-AlGaAs MRR laterally coupled to bus waveguides was demonstrated by Rafizadeh *et al.* in 1997. The members of Ping-Tong Ho's group at the Laboratory for Physical Sciences (LPS), College Park, MD, demonstrated both laterally and vertically coupled rings in GaAs-AlGaAs acting as multi ring devices, switches and routers (Grover *et al.*, 2001a). The GaInAsP-InP material system was problematic for passive microrings because of processing difficulties resulting in high device losses. Nevertheless, the first vertically coupled passive InP-based rings were demonstrated by Ho's group (Grover *et al.*, 2001b; Grover *et al.*, 2002). Other groups have concentrated on disk resonators. The group at the University of Southern California, for example, has demonstrated active and passive vertically coupled microdisk resonators (Djordjev *et al.*, 2002b). From 2008 till now, Yupapin *et al.* used micro ring resonators for several applications such as femtosecond pulse generation, stopping and storing light, entangled photon states generation, THz light pulse generation (Blanchard *et al.*, 2011), photons trapping and medical. In 2010, Pornsuwancharoen *et al.* showed that both fast and slow light can exist in a MRR system.

1.13   HISTORY OF SLOW LIGHT

In 1966, Basov *et al.* showed the propagation of a pulse through a laser amplifier in which the intensity of the pulse was high enough to make a nonlinear optical response

(Schweinsberg *et al.*, 2006). The nonlinear optical saturation of the amplifier gave rise to fast light, an unexpected result since the linear dispersion is normal at the center of an amplifying resonance so that $v_g < c$ is expected for low intensity pulses. The pulse development can be attributed to a nonlinear pulse reshaping effect where the front edge of the pulse depletes the atomic inversion density so that the trailing edge propagates with much lower amplification. In addition, it was found that the effects of dispersion give a negligible contribution to the pulse propagation velocity in comparison to the non-linear optical saturation effects. Such pulse advancement due to amplifier saturation is referred to as super luminous propagation. The propagation of pulses is sufficiently weak so that the linear optical properties of the medium take into consideration. These properties can be modified in a nonlinear fashion by applying an intense auxiliary field.

Icsevgi and Lamb have done a theoretical analysis of the propagation of intense laser pulses through a laser amplifier in 1969 (Schweinsberg *et al.*, 2006). Two types of pulses were distinguished in their numerical solutions it was shown that the pulse with infinite support can propagate with group velocities exceeding that of light in vacuum. For a pulse with compact support, they achieved that the region of the pulse when becoming nonzero cannot propagate faster than the speed of light in vacuum. These results were consistent with the work of Brillouin and a nonlinear optical medium (Bigelow *et al.*, 2003a).

Brillouin in 1914 discussed the distinction between front velocity, group velocity and its implication in the special theory of relativity. These issues have been clarified further in the work of Sherman and Oughstun in 1981. A simple algorithm was used to describe short pulse propagation through dispersive systems in the presence of loss (Sherman & Oughstun, 1981). In 1996, Diener reported that a pulse propagates superluminally, faster than the speed of light. Predictions on fast light was done using an analytic continuation of the pulse that lies within the waveguide (Diener, 1997). In subsequent work, Diener introduced an energy transport velocity as a function of refractive index.

The light velocity less than or equal to the speed of light in vacuum in any material for any value of n. Subsequent experiments were conducted in the late 1960s such as Bieber in 1969 (Boyd & Gauthier, 2002), Frova *et al.* In 1969 (Soares *et al.*, 1995), and in early 1970's by Faxvog *et al.* (Fuji *et al.*, 1997). Slow light can be observed through weak pulses propagating in amplifying media as expected for a linear amplifier. In 1970, Casperson and Yariv showed slow light using a high gain 3.51 μm xenon amplifier (Tovar & Casperson, 1995).

During this same period, Garrett and Mc Cumber in 1970 made an important contribution to this field of fast and slow light via their theoretical investigation regarding the propagation of a weak Gaussian pulse through either an amplifier or absorber. It was shown that the pulse remains substantially Gaussian and unchanged in width for any exponential absorption or gain lengths and the location of the maximum pulse amplitude propagates at v_g, even when $v_g > c$ or $v_g < 0$.

The autocorrelation method was used to measure the pulse shapes sensitive to pulse compression but insensitive to pulse asymmetries oscillations. Katz and Alfano observed significant pulse compression has been observed in pulses (Katz & Alfano, 1982). Chu and Wong explained these phenomena theoretically by the inclusion of higher order dispersion and the group velocity remains a meaningful concept even in the presence of pulse compression (Chu & Wong, 1982).

A typical method to create slow light is the use of Electromagnetically Induced Transparency (EIT). This method was introduced by Vestergaad Hau *et al.* in order to make a material system transparent to resonant laser radiation, while retaining the large and desirable optical properties associated with the material resonant response. In 2001, Bennink *et al.* have predicted slow light effects occurring strongly within materials that consist of a two level atomic structure (Bennink *et al.*, 2001).

In 2002, Longhi *et al.* predicted the great advantages obtained from the light signals speed control within an optical fiber. They demonstrated superluminal optical pulse propagation through fiber Bragg gratings (FBGs) for communication optical systems, (Longhi *et al.*, 2002). Stenner *et al.* used fast light medium that exploits the spectral region of anomalous dispersion between two closely spaced amplifying resonances realized by creating large atomic coherence in a laser driven potassium vapour. He obtained larger pulse advancement for a smooth Gaussian shaped pulse. Thevenaz *et al.* achieved both time advancement and delay in optical fibers using stimulated Brillouin scattering in 2007 (Thévenaz *et al.*, 2007). In 2006, Mok *et al.* obtained considerable delays by launching powerful optical pulses at the edge of the rejection band of the FBG in transmission. The Kerr effect was used to modify the delay via a shift of the FBG (Mok *et al.*, 2006).

Several slow and fast light structures proposed over the last decade. The performances in terms of delay-bandwidth product and efficient tuning mechanisms are still far from commercial applications especially with respect to the complexity of the overall system. The challenge ahead is to achieve a large time delay and advancement tunability over a large bandwidth without dispersion by means of a low-cost and short length device at room temperature (Fan *et al.*, 2005), Brillouin (Okawachi *et al.*, 2005; Song *et al.*, 2005), photonic crystal waveguides (Beggs *et al.*, 2011) and optical fiber (Nasser *et al.*, 2011).

During the period from 2006 to 2008, Luc Thévenaz generated fast and slow light in optical fibers with stimulated Brillouin scattering (SBS) and pump power. In 2009, Jalil *et al.* have illustrated stopping and storing light pulses by using a system consisting an Erbium-Doped Fiber Amplifier (EDFA), a Semiconductor Optical Amplifier (SOA) and a fiber ring resonator (Suhailin *et al.*, 2009). In 2009, Tao Wang experimentally demonstrate continuously tunable pulse propagation in silicon on insulator micro ring resonators with mutual mode coupling, which is induced by nanosized gratings along the ring sidewalls (Wang *et al.*, 2009a).

Recently, Yupapin *et al.* have reported promising results in which the slow, stop and store light can be generated using a soliton pulse traveling within the nonlinear MRRs. The large bandwidth can be compressed coherently with a small group velocity.

Over the last few years, slow light generation is an emerging and very attractive research area of interest. The ability to control the velocity of light is usually referred to slow light (Rybin & Timonen, 2011). Particular potential uses of fast and slow light effects exist within optical communications and optical buffers. In the next chapter, this research expresses the theory of research and introduces parameters and phenomena includes optical soliton, fiber nonlinear, wave dispersion, group velocity and analysis of ring resonators, add-drop system, and PANDA ring resonator. Furthermore, the scattering matrix method is introduced as a mathematical physical tool which relates the input and output power.

Chapter 2

MRR systems and soliton propagating in optical fiber communication

2.1 SOLITON PROPERTIES

The soliton is a self-reinforcing solitary wave that keeps its shape when it travels in a waveguide with constant speed. Solitons can be generated by the cancellation of dispersive and nonlinear effects in a waveguide. Solitons are an important development in the field of optical communications. In terms of intensity, soliton can be classified as dark and bright. Dark soliton is characterized by a localized reduction of intensity with more intense Continuous Wave (CW) background. Bright soliton is characterized as a localized intensity peak above a continuous wave background. Thus dark soliton is considered as a localized intensity dip below a CW background (Tang & Shukla, 2007). In optics, the term soliton is used to refer to an optical field that does not change during propagation because of a delicate balance between nonlinear and linear effects in the medium. In the framework of nonlinear optics, solitons can be further classified as being temporal or spatial, depending on the confinement of light either in time or space during propagation. In the case of spatial solitons the nonlinear effects balance the diffraction and propagate without changing their shape (Saleh *et al.*, 1991). For the temporal solitons, the nonlinear effect balances the dispersion. Therefore the pulses maintain their shape if the electromagnetic fields are already spatially confined (Crutcher *et al.*, 2005).

The propagation of optical fields in fibers is governed by Maxwell's equations given as (Diament, 1990):

$$\nabla \cdot \vec{E} = \frac{\rho}{\varepsilon_0} \tag{2.1}$$

$$\nabla \cdot \vec{B} = 0 \tag{2.2}$$

$$\nabla \times \vec{E} = -\frac{\partial \vec{B}}{\partial t} \tag{2.3}$$

$$\nabla \times \vec{B} = \mu_0 \vec{J} + \mu_0 \varepsilon_0 \frac{\partial \vec{E}}{\partial t} \tag{2.4}$$

where E and B are electric and magnetic flux densities, respectively. J is the current density vector and ρ is the charge density that represents the sources for the electromagnetic field. In the absence of free charges in a medium such as optical fibers, $J = 0$ and $\rho = 0$, where ε_0 is the vacuum permittivity, μ_0 is the vacuum permeability, and P is the induced electric polarizations. Maxwell's equations are used to obtain the wave equation to describe light propagation in optical fibers. By taking the curl of Equation (2.3) and using Equation (2.1), Equation (2.4) can be achieved as:

$$\nabla \times (\nabla \times \vec{E}) = \nabla \times \left(-\frac{\partial \vec{B}}{\partial t} \right) \tag{2.5}$$

$$\nabla(\nabla \cdot \vec{E}) - \nabla^2 \vec{E} = -\frac{\partial}{\partial t}(\nabla \times \vec{B}) \tag{2.6}$$

$$\nabla(\nabla \cdot \vec{E}) - \nabla^2 \vec{E} = -\frac{\partial}{\partial t}\left(\mu_0 \vec{J} + \mu_0 \varepsilon_0 \frac{\partial \vec{E}}{\partial t} \right) \tag{2.7}$$

If we consider $\vec{J} = \frac{\partial \vec{P}}{\partial t}$, the equation can be written as:

$$-\nabla^2 \vec{E} = -\frac{\partial}{\partial t}\left(\mu_0 \frac{\partial \vec{P}}{\partial t} + \mu_0 \varepsilon_0 \frac{\partial \vec{E}}{\partial t} \right) \tag{2.8}$$

Finally, Helmholtz equation can be obtained as:

$$\nabla^2 \vec{E} - \frac{1}{c^2}\frac{\partial^2 \vec{E}}{\partial t^2} = \frac{1}{c^2 \varepsilon_0}\frac{\partial^2 \vec{P}}{\partial t^2}, \tag{2.9}$$

where c is the speed of light in vacuum and the relation $c = 1/\sqrt{(\mu_0 \varepsilon_0)}$ is used. To complete the description, a relation between the induced polarization P and the electric field E is needed. In general, the evaluation of P requires a quantum-mechanical approach. Although such an approach is often necessary when the optical frequency is near to medium resonance, a phenomenological relation is used to relate P and E far from medium resonances. If we include only the third-order nonlinear effects governed by $\chi^{(3)}$, the induced polarization consists of two parts such as:

$$\vec{P} = \vec{P}_L + \vec{P}_{NL} \tag{2.10}$$

where the linear part P_L and the nonlinear part P_{NL} are related to the electric field by the general relations

$$\vec{P}_L = \varepsilon_0 \int_{-\infty}^{+\infty} \chi^1(t - t') \cdot \vec{E}(r, t')dt', \tag{2.11}$$

$$\vec{P}_{NL} = \varepsilon_0 \iiint_{-\infty}^{+\infty} \chi^3(t - t_1, t - t_2, t - t_3) \times \vec{E}(r, t_1)\vec{E}(r, t_2)\vec{E}(r, t_3)dt_1\, dt_2\, dt_3 \tag{2.12}$$

where $\chi^{(1)}$ and $\chi^{(3)}$ are the first- and third-order susceptibility tensors. The second-order nonlinear effects can be neglected due to the medium inversion symmetry. Considerable simplification occurs if the nonlinear response is assumed to be instantaneous so that the time dependence of χ^3 is given by the product of three delta functions $\delta(t - t_1)$. Equations (2.11) and (2.12) reduce to:

$$\vec{P}_L = \varepsilon_0 \chi^1 \cdot \vec{E}, \tag{2.13}$$

$$\vec{P}_{NL} = \varepsilon_0 \chi^3 \vec{E} \cdot \vec{E} \cdot \vec{E}, \tag{2.14}$$

where $\tilde{E}(r, \omega)$ is the Fourier transform of $\vec{E}(r, t)$ defined as:

$$\vec{E}(r, t) = \frac{1}{2\pi} \int_{-\infty}^{+\infty} \tilde{E}(r, \omega) \exp(-i\omega t) d\omega, \tag{2.15}$$

By solving Eq. (2.9) with $P_{NL} = 0$. The Eq. (2.9) is then linear in E and useful to write in the frequency domain as:

$$\nabla^2 \tilde{E} + n(\omega) \frac{\omega^2}{c^2} \tilde{E} = 0 \tag{2.16}$$

The frequency-dependent dielectric constant appears is define as

$$\varepsilon(\omega) = 1 + \tilde{\chi}^1(\omega) + \varepsilon_{NL}, \tag{2.17}$$

where $\tilde{\chi}^1(\omega)$ is the Fourier transform of $\chi^1(t)$.

2.2 EVALUATION OF SOLITON SIGNALS

The Non-Linear Schrödinger Equation (NLSE) is an appropriate equation for describing the propagation of light in optical fibers using normalization parameters such as: the normalized time T_0, the dispersion length L_D and peak power of the pulse P_0. The nonlinear Schrödinger equation in the terms of normalized coordinates can be written as:

$$i\left(\frac{\partial u}{\partial z}\right) - \frac{s}{2}\left(\frac{\partial^2 u}{\partial t^2}\right) + N^2 |u|^2 u + i\left(\frac{\alpha}{2}\right) u = 0 \tag{2.18}$$

where, $u(z, t)$ is pulse envelope function, z is propagation distance along the fiber, N is an integer designating the order of soliton and α is the coefficient of energy gain per unit length, and with negative value it represents energy loss. Here, s is -1 for negative β_2 (anomalous GVD-Bright soliton) and $+1$ for positive β_2 (normal GVD-Dark soliton) as shown in Figures 2.1 and 2.2,

$$N_2 = \frac{L_D}{L_{NL}} = \frac{\gamma P_0 T_0^2}{|\beta_2|^2} \tag{2.19}$$

With nonlinear parameter γ and nonlinear length L_{NL}.

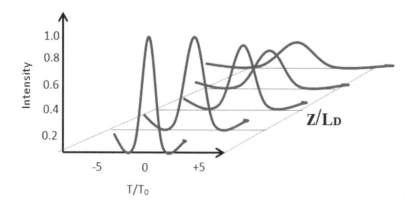

Figure 2.1 Evolution of soliton in normal dispersion regime.

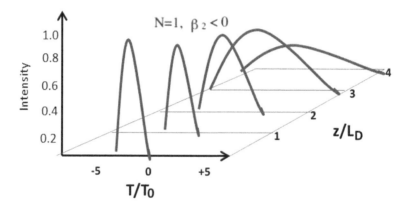

Figure 2.2 Evolution of soliton in anomalous dispersion regime.

It is apparent that SPM dominates for $N > 1$ while for $N < 1$ dispersion effects dominates. For $N \approx 1$ both SPM and GVD cooperate in such a way that the SPM-induced chirp is just right to cancel the GVD induced broadening of the pulse. The optical pulse would then propagate undistorted in the form of soliton. By integrating the NLS, the solution for the fundamental soliton can be written as

$$u(z,t) = \text{sech}(t)\exp(iz/2) \tag{2.20}$$

where, sech(t) is hyperbolic scent function. Since the phase term exp($iz/2$) has no influence on the shape of the pulse, the soliton is independent of z and hence is non dispersive in time domain. It is the property of a fundamental soliton that makes it an ideal candidate for optical communications. Optical solitons are very stable against perturbations; therefore they can be created even when the pulse shape and peak power deviates from ideal conditions (values corresponding to $N = 1$). To have the secured communication, the performance of resonators should be considered in terms

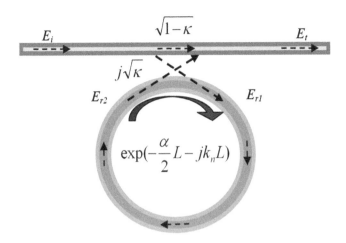

Figure 2.3 Schematic diagram for a ring resonator coupled to a single waveguide.

of resonance width, the free spectral range, the finesse, and the quality factor (Ridha *et al.*, 2010a). A fiber optic ring resonator consists of a waveguide in a closed loop which is coupled to one or more input/output (or bus) waveguides. A simple MRR is shown in Figure 2.3.

2.3 MRR USED TO GENERATE CHAOTIC SIGNALS

A ring resonator provides traveling wave procedure, unlike the standing wave characteristic of Fabry-Perot resonators (F-P) (Ali *et al.*, 2011; Amiri *et al.*, 2015b). A ring resonator can be considered as an interferometer device, which resonates for light whose phase change is an integer multiple of 2π after each trip around the ring. The part of light that does not contribute to this resonant condition will be transmitted through the bus waveguide. **Signal loss** occurs when light is transmitted through the fiber, especially over long distances such as undersea cables. The expression for the resonant wavelengths of the ring is very similar to that of the F-P and is given by

$$\lambda_r = \frac{2\pi R n_{\text{eff}}}{m} \tag{2.21}$$

where R is the ring radius constructed with circular waveguide and m is an integer. In this situation the device will act as a phase filter where all wavelengths are transmitted and the resonant wavelengths, having also traversed the ring, acquire a phase change. To capture or separate the resonant wavelengths from the rest, an additional waveguide as an output bus can be positioned on the opposite side of the ring (Ali *et al.*, 2010g; Amiri *et al.*, 2014b). In this case the ring resonator is known as an add/drop filter system. The key performance parameters of the ring resonator include the free spectral range (FSR), the extinction ratio (ER), and the finesse. The expression for the FSR of a ring resonator is given by

$$\Delta\lambda = \frac{\lambda_r^2}{2\pi R n_g} \tag{2.22}$$

The nonlinearity of the fiber ring is of the Kerr-type, wherein the nonlinear refractive index is given by Ali *et al.* (2010h) and Saktioto *et al.* (2010a)

$$n = n_0 + n_2 I = n_0 + \left(\frac{n_2}{A_{\text{eff}}}\right) P, \tag{2.23}$$

where n_0 and n_2 are the linear and nonlinear refractive indices, while I and P are the optical intensity and optical field power, respectively. Here, the fiber coupler is considered as a point device and is reciprocal. The linear and nonlinear phase shifts of the ring resonator can be expressed by $\phi_0 = kLn_0$ and $\phi_{NL} = kLn_2|E_1|^2$, where $k = 2\pi/\lambda$ is a wave number, and $L = 2\pi R$ is the circumference of the ring resonator, where R is the radius of the ring resonator (Amiri *et al.*, 2012c; Alavi *et al.*, 2014b). Mathematically, the subsequence equations of the round-trip within the system is given by

$$E_{n+1} = j\sqrt{(1-\gamma)\kappa} E_{\text{in}} + \sqrt{(1-\gamma)(1-\kappa)} x E_n \exp(-j(\phi_0 + \phi_{NL})) \tag{2.24}$$

Here, the subscript n denotes the number of round-trips inside the system. This equation has to be satisfied with boundary conditions appropriate for a ring. The transmission around the single ring resonator is represented by

$$z^{-1} = \exp(-\alpha L/2 - jk_n L) \tag{2.25}$$

where k_n is the propagation constant and $\alpha L/2$ is the ring loss (round-trip loss), which includes propagation loss, losses resulting from transitions in the curvature, and bending losses. The value of α (unit length^{-1}) depends on the properties of the material and the waveguide used, and it is referred to as the intensity attenuation coefficient, where L is the circumference of the ring resonator (Gifany *et al.*, 2013; Amiri *et al.*, 2014g; Alavi *et al.*, 2015d). In order to describe this, we consider a ring resonator connected to a single coupler that extracts light from the ring into the output waveguides.

When an input electric field E_i is coupled to the ring waveguide through an external bus waveguide, a positive feedback is induced and the field inside the ring resonator E_{r2} starts to build up. The feedback mechanism will be induced by the ring waveguide, and therefore does not have any further requirements such as Bragg gratings, mirrors, or distributed feedback waveguides with difficult fabrication process. Due to on-resonant certain wavelength of the input signals inside the ring waveguide, frequency selectivity is obtained (Glaser, 1997). The inserted and transmitted electric fields into the ring resonator are expressed by

$$E_{r1} = (1-\gamma)^{\frac{1}{2}} \left[jE_i\sqrt{\kappa} + E_{r2}\sqrt{1-\kappa} \right] \tag{2.26}$$

$$E_{r2} = E_{r1} \exp\left(-\frac{\alpha}{2}L - jk_nL\right) \tag{2.27}$$

where $k_n = \frac{2\pi \cdot n_{\text{eff}}}{\lambda}$ and γ denotes the intensity insertion loss coefficient of the directional coupler and n_{eff} is the effective refractive index. Therefore, the refractive index n quantifies the increase in the wave number (phase change per unit length) caused by the medium. Here, the effective refractive index n_{eff} has the similar meaning with light propagation in a waveguide, where it depends not only on the wavelength but also on the mode, in which the light propagates. The ratio of the output and input powers which is E_t/E_i can be calculated as

$$\frac{E_t}{E_i} = (1-\gamma)^{\frac{1}{2}} \cdot \left[\frac{\sqrt{1-\kappa} - (1-\gamma)^{\frac{1}{2}} \cdot \exp(-\frac{\alpha}{2}L - jk_nL)}{1 - (1-\gamma)^{\frac{1}{2}} \cdot \sqrt{1-\kappa} \cdot \exp(-\frac{\alpha}{2}L - jk_nL)}\right] \tag{2.28}$$

In the following a new parameter will be used for simplifying:

$$D = (1-\gamma)^{\frac{1}{2}}, \quad x = D \cdot \exp\left(-\frac{\alpha}{2} \cdot L\right), \quad y = \sqrt{1-\kappa}, \quad \phi = k_nL$$

Intensity relation to the output port is given by Okamoto (2006):

$$T = \frac{I_t}{I_i}(\varphi) = \left|\frac{E_t}{E_i}\right|^2 = D^2 \cdot \left[1 - \frac{(1-x^2) \cdot (1-y^2)}{(1-xy)^2 + 4xy \cdot \sin^2(\frac{\varphi}{2})}\right] \tag{2.29}$$

Maximum and minimum transmission can be calculated when $\sin^2\left(\frac{\varphi}{2}\right)$ is "1" and "0" respectively. Therefore;

$$T_{\text{max}} = D^2 \cdot \frac{(x+y)^2}{(1+x\cdot y)^2} \tag{2.30}$$

$$T_{\text{min}} = D^2 \cdot \frac{(x-y)^2}{(1-x\cdot y)^2} \tag{2.31}$$

The minimum transmission, T_{min} occurs at the resonant point when the circumference of the ring L, is an integer number of the guide wavelength, which is given by

$$\phi = k_n \cdot L = 2m\pi, \quad m = \text{integer},$$

$$m \cdot \lambda_m = n \cdot L \tag{2.32}$$

Here, m is the mode number, λ_m is the resonant mode wavelength. The on-off ratio for the single ring resonator is defined as the ratio of the on-resonance intensity to the off-resonance intensity which is maximum at $T_{\text{min}} = 0$. Therefore $x = y$ and

$$\alpha = -\frac{1}{L} \times \ln\left(\frac{1-\kappa}{D^2}\right) \tag{2.33}$$

This relationship given by equation (3.26) is also referred to as critical coupling, where the maximum on-off ratio $\frac{I_t}{I_i}(2m\pi) = 0$ can be obtained by varying the coupling coefficient (κ) or the intensity attenuation coefficient (α).

2.4 RESONANCE BANDWIDTH OF SOLITON

Resonance bandwidth determines how fast optical data can be processed by a ring resonator. The resonator bandwidth is given by the full-width at half-maximum (FWHM or 3 dB bandwidth) $\delta\phi[I_t/I_i(\varphi) = 0.5]$ and the finesse F of the resonator is given by:

$$\delta\phi = \frac{2(1 - xy)}{\sqrt{xy}} \tag{2.34}$$

To understand how the bandwidth of the resonator is affected by the coupling coefficient κ, we will consider a critically coupled ring resonator. In such a case,

$$\delta\phi = \frac{2\kappa}{\sqrt{1 - \kappa}} \tag{2.35}$$

Therefore, the lower coupling coefficient, the smaller resonance bandwidth is obtained.

2.5 FINESSE OF SOLITON

The finesse of the resonator is defined as a ratio of the free spectral range and the full width at half maximum of the resonance. For the Figure 2.4 using FSR (frequency spacing between two resonances) in terms of the is equal to 2π and thus the finesse is given by Nikoukar *et al.* (2013) and Amiri & Ali (2014a)

$$F = \frac{2\pi}{\delta\phi} = \frac{\pi\sqrt{xy}}{(1 - xy)} \tag{2.36}$$

2.6 FREE SPECTRAL RANGE (FSR) OF SOLITON

The frequency spacing between two resonance peaks is called the free spectral range which can be calculated. The phase constant which corresponds to $\phi = 2(m + 1)\pi$ is defined as κ. The phase constant corresponds to $\phi = 2(m + 1)\pi$ is defined as $\kappa + \Delta\kappa$. The frequency shift Δf and the wavelength shift $\Delta\lambda$ are related to the variation of the phase constant $\Delta\kappa$ as $\Delta f = (c/2\pi) \cdot \Delta\kappa$ and $\Delta\lambda = -(\lambda^2/2\pi) \cdot \Delta\kappa$. The resonance spacing in terms of the frequency f and the wavelength λ are given by

$$\Delta f = \frac{c}{n_{gr} \cdot L} \tag{2.37}$$

$$\Delta\lambda = \left| -\frac{\lambda^2}{n_{gr} \cdot L} \right| \tag{2.38}$$

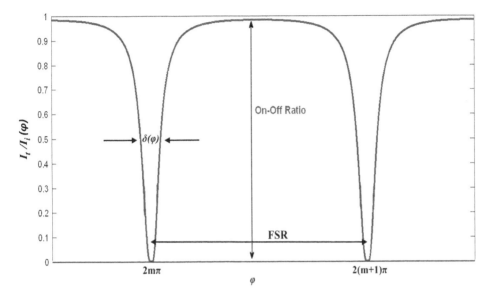

Figure 2.4 Transmission characteristic of single ring resonator.

where n_{gr} is the group refractive index, which is defined as;

$$n_{gr} = n_{\text{eff}} - \lambda \frac{dn_{\text{eff}}}{d\lambda}\bigg| \qquad (2.39)$$

2.7   QUALITY FACTOR OF SOLITON

Another value for characterization of ring resonator is the Q factor, The Q factor of the resonator is a measure of the sharpness of the resonance. In analogy with electrical circuit, the quality factor of an optical waveguide due it stored energy and the power lost per optical cycle. The Q factor is defined as (Amiri & Ali, 2013d; Parisa Naraei *et al.*, 2014)

$$Q = \omega \frac{\text{stored energy}}{\text{Power Loss}} \qquad (2.40)$$

where ω is the frequency of the light coupled to the resonator. The Q factor of the resonator can be calculated from

$$Q = \frac{f_0}{\delta f} = \frac{\lambda_0}{\delta \lambda} \qquad (2.41)$$

The Q factor is the ratio of the absolute frequency f_0 or absolute wavelength λ_0 to the 3 dB bandwidth (δf or $\delta \lambda$). The shape and the bandwidth of the fiber response is determined by Q factor. The finesse and the Q factor are both important when one

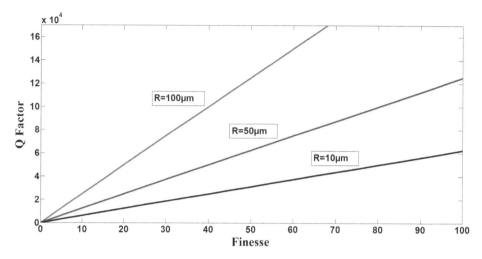

Figure 2.5 Q-Factor depending on the Finesse for a specific radius R.

is interested in both the FSR (Δf or $\Delta\lambda$) and the 3 dB bandwidth (δf or $\delta\lambda$). They are related by:

$$\frac{Q}{F} = \frac{f_0}{\Delta f} = \frac{\lambda_0}{\Delta\lambda} \tag{2.42}$$

The Q factor depending on the finesse F for a ring resonator with a radius $R = 100\,\mu\text{m}$, $50\,\mu\text{m}$ and $10\,\mu\text{m}$, a group refractive index $n_{gr} = 3.44$ at a wavelength of $\lambda = 1.55\,\mu\text{m}$ is shown in Figure 2.5.

2.8 CHAOTIC SOLITON SIGNAL GENERATOR

An add/drop ring resonator configuration connected to a single ring resonator depicted in Figure 2.6, is constructed by the fiber optic using optical couplers, where the circumference of the fiber ring is L. Here, the input pulse to the ring resonator is given by $E_{\text{in}}(t)$, where the output signal is expressed by $E_{\text{out}}(t)$.

The input light is a monochromatic laser pulse with constant amplitude and random phase modulation, which results in temporal coherence degradation. It can be expressed as

$$E_{\text{in}}(t) = E_0 \exp\left[\left(\frac{z}{2L_D}\right) - i\omega_0 t\right] \tag{2.43}$$

E_0 and z are the amplitude of optical field and propagation distance respectively (Amiri *et al.*, 2014c; Amiri & Naraei, 2014). L_D is the dispersion length of the soliton pulse where frequency shift of the signal is ω_0. When a soliton pulse is input and propagated within a MRR as shown in Figure 2.6, the normalized output of the light field is defined

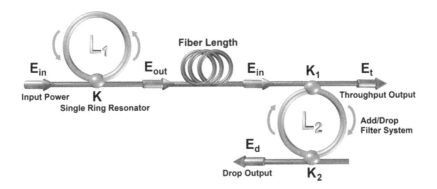

Figure 2.6 A fiber optic ring resonator is constructed to an add/drop filter system by the couplers.

as the ratio between the output and input fields $E_{out}(t)$ and $E_{in}(t)$ respectively in each round-trip and it can be expressed as (Sanati *et al.*, 2014),

$$\left|\frac{E_{out}(t)}{E_{in}(t)}\right|^2 = (1-\gamma)\left[1 - \frac{(1-(1-\gamma)x^2)\kappa}{(1-x\sqrt{1-\gamma}\sqrt{1-\kappa})^2 + 4x\sqrt{1-\gamma}\sqrt{1-\kappa}\sin^2\left(\frac{\phi}{2}\right)}\right] \quad (2.44)$$

This system is very similar to a Fabry-Perot cavity, which has an input and output mirror with a field reflectivity, $(1-\kappa)$, and a fully reflecting mirror. κ is the coupling coefficient, and $x = \exp(-\alpha L/2)$ represents a round-trip loss coefficient, $\phi = \phi_0 + \phi_{NL}$, where $\phi_0 = kLn_0$ and $\phi_{NL} = kLn_2|E_{in}|^2$ are the linear and nonlinear phase shifts, $k = 2\pi/\lambda$ is the wave propagation number in a vacuum. Here, L and α are a waveguide length and linear absorption coefficient, respectively (Ali *et al.*, 2010b; Bahadoran *et al.*, 2011; Amiri & Ali, 2012).

2.9 ADD/DROP FILTER SYSTEM

Recently, optical ring resonators (ORR) have numerous applications in single mode lasers, biosensors, optical switching, add/drop filters, tunable lasers, signal processing and dispersion compensators (Kouhnavard *et al.*, 2010b; Amiri *et al.*, 2011b; Shahidinejad *et al.*, 2014; Amiri *et al.*, 2015). In any WDM system, optical filters are used for separating one optical channel from the combined signals. The basic ORR with two couplers is illustrated in Figure 2.7. The main performance characteristics of these resonators are the transmittance, free spectral range, finesse, Q-factor, and the group delay, which have been demonstrated both theoretically and experimentally in many works. Structural design of a single ring resonator (SRR) add/drop filter system is shown in Figure 2.7, which is constructed by 2×2 optical couplers.

For simplification, the intensity relation (Yariv, 2000) does not take into account coupling losses ($D^2 = 1$)

$$E_a = E_{i1}j\sqrt{\kappa_1} + E_b\sqrt{1-\kappa_1}e^{\frac{-\alpha}{2}\frac{L}{2}-jkn\frac{L}{2}} \quad (2.45)$$

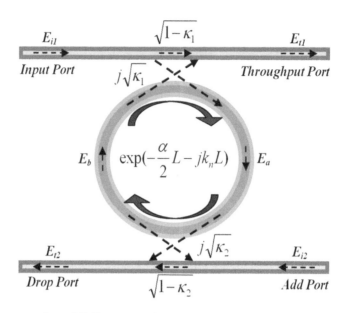

Figure 2.7   Ring resonator with two adjacent waveguide.

$$E_b = E_a\sqrt{1-\kappa_2}\, e^{\frac{-\alpha}{2}\frac{L}{2}-jk_n\frac{L}{2}} \tag{2.46}$$

$$E_a = \frac{E_{i1}j\sqrt{\kappa_1}}{1-\sqrt{1-\kappa_1}\sqrt{1-\kappa_2}\, e^{\frac{-\alpha}{2}L-jk_nL}} \tag{2.47}$$

$$E_b = \frac{E_{i1}j\sqrt{\kappa_1}}{1-\sqrt{1-\kappa_1}\sqrt{1-\kappa_2}\, e^{\frac{-\alpha}{2}L-jk_nL}} \cdot \sqrt{1-\kappa_2}\, e^{\frac{-\alpha}{2}\frac{L}{2}-jk_n\frac{L}{2}} \tag{2.48}$$

$$E_{t1} = E_b e^{\frac{-\alpha}{2}\frac{L}{2}-jk_n\frac{L}{2}} j\sqrt{\kappa_1} + E_{i1}\sqrt{1-\kappa_1} \tag{2.49}$$

$$E_{t2} = E_a e^{\frac{-\alpha}{2}\frac{L}{2}-jk_n\frac{L}{2}} j\sqrt{\kappa_2} \quad \text{at} \quad E_{i2}=0 \tag{2.50}$$

$$\frac{E_{t1}}{E_{i1}} = \frac{-\kappa_1\sqrt{1-\kappa_2}\, e^{\frac{-\alpha}{2}L-jk_nL} + \sqrt{1-\kappa_1} - (1-\kappa_1)\sqrt{1-\kappa_2}\, e^{\frac{-\alpha}{2}L-jk_nL}}{1-\sqrt{1-\kappa_1}\sqrt{1-\kappa_2}\, e^{\frac{-\alpha}{2}L-jk_nL}}$$

$$= \frac{-\sqrt{1-\kappa_2}\, e^{\frac{-\alpha}{2}L-jk_nL} + \sqrt{1-\kappa_1}}{1-\sqrt{1-\kappa_1}\sqrt{1-\kappa_2}\, e^{\frac{-\alpha}{2}L-jk_nL}} \tag{2.51}$$

$$\frac{E_{t2}}{E_{i1}} = \frac{-\sqrt{\kappa_1\cdot\kappa_2}\, e^{\frac{-\alpha}{2}\frac{L}{2}-jk_n\frac{L}{2}}}{1-\sqrt{1-\kappa_1}\sqrt{1-\kappa_2}\, e^{\frac{-\alpha}{2}L-jk_nL}} \tag{2.52}$$

where κ_1 and κ_2 are the coupling coefficients, $L=2\pi R$ and R is the radius of the add/drop filter device (Amiri *et al.*, 2012d; Amiri & Shahidinejad, 2014). The normalized outputs of the add/drop filter system are expressed as (Amiri *et al.*, 2015a; Alavi *et al.*, 2015c):

$$\frac{I_{t1}}{I_{i1}} = \left|\frac{E_{t1}}{E_{i1}}\right|^2 = \frac{1 - \kappa_1 - 2\sqrt{1-\kappa_1}\sqrt{1-\kappa_2}e^{\frac{-\alpha}{2}L}\cos(k_nL) + (1-\kappa_2)e^{-\alpha L}}{1 + (1-\kappa_1)(1-\kappa_2)e^{-\alpha L} - 2\sqrt{1-\kappa_1}\sqrt{1-\kappa_2}e^{\frac{-\alpha}{2}L}\cos(k_nL)} \quad (2.53)$$

$$\frac{I_{t2}}{I_{i1}} = \left|\frac{E_{t2}}{E_{i1}}\right|^2 = \frac{\kappa_1 \cdot \kappa_2 e^{\frac{-\alpha}{2}L}}{1 + (1-\kappa_1)(1-\kappa_2)e^{-\alpha L} - 2\sqrt{1-\kappa_1}\sqrt{1-\kappa_2}e^{\frac{-\alpha}{2}L}\cos(k_nL)} \quad (2.54)$$

Using $y_1 = \sqrt{1-\kappa_1}$ and $y_2 = \sqrt{1-\kappa_2}$, the intensity relations are then given by:

$$\frac{I_{t1}}{I_{i1}}(\varphi) = \left|\frac{E_{t1}}{E_{i1}}\right|^2 = 1 - \frac{(1-y_1^2)\cdot(1-y_2^2x^2)}{(1-y_1y_2x)^2 + 4y_1y_2x\sin^2\left(\frac{\varphi}{2}\right)} \quad (2.55)$$

$$\frac{I_{t2}}{I_{i1}}(\varphi) = \left|\frac{E_{t2}}{E_{i1}}\right|^2 = \frac{(1-y_1^2)\cdot(1-y_2^2)\cdot x}{(1-y_1y_2x)^2 + 4y_1y_2x\sin^2\left(\frac{\varphi}{2}\right)} \quad (2.56)$$

The full-width at half-maximum (FWHM) is given in this configuration by:

$$\delta\phi = 2\frac{1-y_1y_2x}{\sqrt{y_1y_2x}}, \quad (2.57)$$

where the finesse F is given by:

$$F = \frac{2\pi}{\delta\phi} = \frac{\pi\sqrt{y_1y_2x}}{1-y_1y_2x} \quad (2.58)$$

The maximum and minimum transmission are calculated as follows. For the throughput port:

$$T_{\max} = \frac{(y_1 + y_2x)^2}{(1+y_1y_2x)^2} \quad (2.59)$$

$$T_{\min} = \frac{(y_1 - y_2x)^2}{(1-y_1y_2x)^2} \quad (2.60)$$

And for the drop port:

$$T_{\max} = \frac{(1-y_1^2)\cdot(1-y_2^2)\cdot x}{(1-y_1y_2x)^2} \quad (2.61)$$

$$T_{\min} = \frac{(1-y_1^2)\cdot(1-y_2^2)\cdot x}{(1+y_1y_2x)^2} \quad (2.62)$$

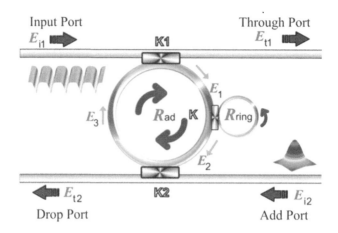

Input Port   Through Port

Figure 2.8 A schematic diagram of Half-Panda system.

The on-off ratio of an add/drop filter system is given by:

$$\frac{T_{\max}(\text{through put port})}{T_{\min}(\text{drop port})} = \frac{(y_1 + y_2 x)^2}{(1 - y_1^2) \cdot (1 - y_2^2) \cdot x} \tag{2.63}$$

The output intensity, I_{t1} at the throughput port will be zero at resonance ($k_n L = 2m\pi$) which indicates that the resonance wavelength is fully extracted by the resonator when $\kappa_1 = \kappa_2$ and $\alpha = 0$. The loss of signal power resulting from the insertion of a device in a transmission line for example an optical fiber is defined insertion loss and usually expressed in dBs. Therefore, it is a measure of attenuation. Attenuation can include loss due to the source and load impedances not matching, but is not included in insertion loss since this is a loss that was already present before the "insertion" was made. If the power transmitted to the load before insertion is P_T and the power received by the load after insertion is P_R, then the insertion loss in *dB* is given by

$$IL = 10 \log_{10} \frac{P_T}{P_R} \tag{2.64}$$

2.10 HALF PANDA RING RESONATOR FUNCTION

The dark soliton pulse is introduced into the multiplexer half-Panda system shown in Figure 2.8. This system consists of an add-drop ring system connected to a smaller ring resonator on the right side. The dynamic behavior of the optical tweezers is appeared when the bright soliton is inputted into the add port of the system. The ring resonator with a radius (R_{ring}) of $10\,\mu$m and a coupling coefficient (κ) of $\kappa = 0.2$ is connected to the add-drop system with radius (R_{ad}) of $15\,\mu$m and coupling coefficient of $\kappa_1 = \kappa_2 = 0.3$. The effective area of the coupling section is $A_{\text{eff}} = 25\,\mu\text{m}^2$.

The input optical field (E_{i1}) of the dark soliton and added optical field (E_{i2}) of the bright soliton are given by Afroozeh et al. (2014b), Amiri & Ali (2014c) and Amiri et al. (2015a)

$$E_{i1} = A \tanh\left[\frac{T}{T_0}\right] \exp\left[\left(\frac{z}{2L_D}\right) - i\omega_0 t\right] \tag{2.65}$$

$$E_{i2} = A \operatorname{sech}\left[\frac{T}{T_0}\right] \exp\left[\left(\frac{z}{2L_D}\right) - i\omega_0 t\right] \tag{2.66}$$

A and z are the optical field amplitude and propagation distance, respectively. T is defined as soliton pulse propagation time in a frame moving at the group velocity, $T = t - \beta_1 \times z$, where β_1 and β_2 are the coefficients of the linear and second order terms of the Taylor expansion of the propagation constant. $L_D = T_0^2/|\beta_2|$ represents the dispersion length of the soliton pulse, where the carrier frequency of the soliton is ω_0. When a soliton pulse keeps its temporal width invariance as it propagates, it is called a temporal soliton. T_0 is known for the intensity of soliton peak as $(|\beta_2/\Gamma T_0^2|)$. A balance should be achieved between the dispersion length (L_D) and the nonlinear length $L_{NL} = (1/\gamma\varphi_{NL})$, where γ and φ_{NL} are the coupling loss of the field amplitude and nonlinear phase shift, thus $L_{NL} = L_D$ should be satisfied. Within the nonlinear medium, the refractive index (n) changes (Amiri & Ali, 2014b; Amiri et al., 2014b; Amiri et al., 2015d):

$$n = n_0 + n_2 I = n_0 + \left(\frac{n_2}{A_{\text{eff}}}\right) P, \tag{2.67}$$

n_0 and n_2 are the linear and nonlinear refractive indexes, respectively. I and P represent the optical intensity and optical power, respectively. The effective mode core area of the device is given by A_{eff}. The output and input signals in each round-trip of the ring on the right side can be calculated (Afroozeh et al., 2010a; Ali et al., 2010q; Amiri, 2011a):

$$\left|\frac{E_2}{E_1}\right|^2 = (1 - \gamma)\left[1 - \frac{(1 - (1 - \gamma)x^2)\kappa}{(1 - x\sqrt{1 - \gamma}\sqrt{1 - \kappa})^2 + 4x\sqrt{1 - \gamma}\sqrt{1 - \kappa}\sin^2\left(\frac{\phi}{2}\right)}\right] \tag{2.68}$$

Here, the $E_1(t)$ is the electric field inserted into the ring resonator, where the output signal is shown by $E_2(t)$. Therefore ring resonator can be comparable to a Fabry-Perot cavity. It has an input and an output mirror with a field reflectivity, $(1 - \kappa)$, and a fully reflecting mirror. Here κ is the coupling coefficient, and $x = \exp(-\alpha L/2)$ represents a round-trip loss coefficient, $\varphi_0 = kLn_0$ and $\varphi_{NL} = kLn_2|E_{in}|^2$ are the linear and nonlinear phase shifts, $k = 2\pi/\lambda$ is the wave propagation number in a vacuum (Ali et al., 2010c; Shojaei & Amiri, 2011b; Afroozeh et al., 2012b). L and α are a waveguide length and linear absorption coefficient, respectively. In this work, the iterative method is inserted to obtain the required results using Equation (2.68). The electric field of the left side of the add-drop ring resonator can be expressed by Equation (2.69):

$$E_3 = \sqrt{1 - \gamma_2} \times \left[E_2 \times \sqrt{1 - \kappa_2} + j\sqrt{\kappa_2} \times E_{i2}\right] \tag{2.69}$$

Here we define the E_0 which can be expressed by

$$E_0 = E_1 \frac{\sqrt{1 - \gamma(1 - \kappa)} - (1 - \gamma)e^{-\frac{\alpha}{2}L_{\text{ring}} - jk_n L_{\text{ring}}}}{1 - \sqrt{1 - \gamma}\sqrt{1 - \kappa}e^{-\frac{\alpha}{2}L_{\text{ring}} - jk_n L_{\text{ring}}}}, \tag{2.70}$$

and it is the electric field of the small ring on the right side of the Half-Panda system. The output fields, E_{t1} and E_{t2} at the throughput and drop parts of the Half-Panda are expressed by

$$E_{t1} = -x_1 x_2 y_2 \sqrt{\kappa_1} E_{i2} e^{-\frac{\alpha L_{\text{ad}}}{2}} - jk_n \frac{L_{\text{ad}}}{2}$$

$$+ \left[\frac{x_2 x_3 \kappa_1 \sqrt{\kappa_2} E_0 E_{i1}\left(e^{-\frac{\alpha L_{\text{ad}}}{2}} - jk_n \frac{L_{\text{ad}}}{2}\right)^2 + x_3 x_4 y_1 y_2 \sqrt{\kappa_1}\sqrt{\kappa_2} E_0 E_{i2}\left(e^{-\frac{\alpha L_{\text{ad}}}{2}} - jk_n \frac{L_{\text{ad}}}{2}\right)^3}{1 - x_1 x_2 y_1 y_2 E_0 \left(e^{-\frac{\alpha L_{\text{ad}}}{2}} - jk_n \frac{L_{\text{ad}}}{2}\right)^2} \right] \tag{2.71}$$

$$E_{t2} = x_2 y_2 E_{i2}$$

$$+ \left[\frac{x_1 x_2 \kappa_1 \sqrt{\kappa_1}\sqrt{\kappa_2} E_0 E_{i1} e^{-\frac{\alpha L_{\text{ad}}}{2}} - jk_n \frac{L_{\text{ad}}}{2} + x_1 x_3 y_1 y_2 \sqrt{\kappa_2} E_0 E_{i2}\left(e^{-\frac{\alpha L_{\text{ad}}}{2}} - jk_n \frac{L_{\text{ad}}}{2}\right)^2}{1 - x_1 x_2 y_1 y_2 E_0 \left(e^{-\frac{\alpha L_{\text{ad}}}{2}} - jk_n \frac{L_{\text{ad}}}{2}\right)^2} \right]. \tag{2.72}$$

Here, we define $x_1 = \sqrt{1 - \gamma_1}$, $x_2 = \sqrt{1 - \gamma_2}$, $x_3 = 1 - \gamma_1$, $x_4 = 1 - \gamma_2$, $y_1 = \sqrt{1 - \kappa_1}$ and $y_2 = \sqrt{1 - \kappa_2} \cdot E_{t1}$ and E_{t2} represent the optical fields of the throughput and drop ports respectively (Afroozeh et al., 2010b; Amiri et al., 2011d; Amiri et al., 2014a). $L_{\text{ad}} = 2\pi R_{\text{ad}}$, where R_{ad} is the radius of the ring. The waveguide (ring resonator) loss is $\alpha = 0.1\,\text{dBmm}^{-1}$. The fractional coupler intensity loss is $\gamma = 0.1$, where $L_{\text{ring}} = 2\pi R_{\text{ring}}$ and R_{ring} is the radius of the ring. The chaotic noise cancellation can be managed by using the specific parameters of the add-drop system in which required signals can be retrieved by the specific users.

2.11 PANDA RING RESONATORS

This system consists of one add/drop interferometer system connected to two ring resonators in the left and right sides. This system represents a new technique of combination and integration of MRRs in which it can be widely used to improve the secure communication and the high capacity of optical signal proceeding in network communications (Amiri et al., 2013; Amiri et al., 2015c). Here the derived equations of the system is introduced which show that how does the input pulse propagates inside the rings systems. The proposed system is shown in Figure 2.9.

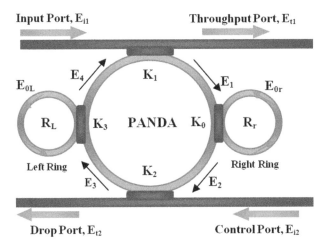

Figure 2.9 Schematic of a PANDA ring resonator system.

The resonator output fields, E_{t1} and E_1 consist of the transmitted and circulated components within the add/drop optical filter system, given by

$$E_{t1} = \sqrt{1-\gamma_1}\left[\sqrt{1-\kappa_1}E_{i1} + j\sqrt{\kappa_1}E_4\right] \tag{2.73}$$

$$E_1 = \sqrt{1-\gamma_1}\left[\sqrt{1-\kappa_1}E_4 + j\sqrt{\kappa_1}E_{i1}\right] \tag{2.74}$$

$$E_2 = E_{0r}E_1 e^{-\frac{\alpha}{2}\frac{L}{2}-jk_n\frac{L}{2}} \tag{2.75}$$

where κ_1 is the intensity coupling coefficient, γ_1 is the fractional coupler intensity loss, α is the attenuation coefficient, $k_n = \frac{2\pi}{\lambda}$ is the wave propagation number, λ is the input wavelength light field, $L = 2\pi R_{ad}$ and R_{ad} is the radius of the add/drop system. For the second coupler of the add/drop system (Amiri *et al.*, 2013d; Alavi *et al.*, 2013b)

$$E_{t2} = \sqrt{1-\gamma_2}\left[\sqrt{1-\kappa_2}E_{i2} + j\sqrt{\kappa_2}E_2\right] \tag{2.76}$$

$$E_3 = \sqrt{1-\gamma_2}\left[\sqrt{1-\kappa_2}E_2 + j\sqrt{\kappa_2}E_{i2}\right] \tag{2.77}$$

$$E_4 = E_{0L}E_3 e^{-\frac{\alpha}{2}\frac{L}{2}-jk_n\frac{L}{2}} \tag{2.78}$$

E_{0r} and E_{0L} are the light fields circulated components of the nano-ring radii and R_r and R_L are the coupled rings into the right and left sides of the add/drop optical filter system, respectively. Transmitted and circulated components of the light fields in the right nano-ring, R_r are given by Afroozeh *et al.* (2010c), Amiri *et al.* (2014b) and Neo *et al.* (2014)

$$E_2 = \sqrt{1-\gamma}\left[\sqrt{1-\kappa_0}E_1 + j\sqrt{\kappa_0}E_{r2}\right] \tag{2.79}$$

$$E_{r1} = \sqrt{1-\gamma}\left[\sqrt{1-\kappa_0}E_{r2} + j\sqrt{\kappa_0}E_1\right] \tag{2.80}$$

$$E_{r2} = E_{r1}e^{-\frac{\alpha}{2}L_1 - jk_nL_1}. \tag{2.81}$$

or

$$E_{r1} = \frac{j\sqrt{1-\gamma}\sqrt{\kappa_0}E_1}{1 - \sqrt{1-\gamma}\sqrt{1-\kappa_0}e^{-\frac{\alpha}{2}L_1 - jk_nL_1}} \tag{2.82}$$

$$E_{r2} = \frac{j\sqrt{1-\gamma}\sqrt{\kappa_0}E_1 e^{-\frac{\alpha}{2}L_1 - jk_nL_1}}{1 - \sqrt{1-\gamma}\sqrt{1-\kappa_0}e^{-\frac{\alpha}{2}L_1 - jk_nL_1}} \tag{2.83}$$

where $L_1 = 2\pi R_r$ and R_r is the radius of the right side nano-ring. Thus, the output circulated light field, E_{0r}, for the right side nano-ring is given by

$$E_{0r} = E_1 \frac{\sqrt{(1-\gamma)(1-\kappa_0)} - (1-\gamma)e^{-\frac{\alpha}{2}L_1 - jk_nL_1}}{1 - \sqrt{1-\gamma}\sqrt{1-\kappa_0}e^{-\frac{\alpha}{2}L_1 - jk_nL_1}} \tag{2.84}$$

Similarly, the output circulated light field, E_{0L}, for the left side nanoring of the add/drop system is given by (Ali *et al.*, 2010f; Amiri *et al.*, 2012f; Amiri *et al.*, 2015b)

$$E_{0L} = E_3 \frac{\sqrt{(1-\gamma_3)(1-\kappa_3)} - (1-\gamma_3)e^{-\frac{\alpha}{2}L_2 - jk_nL_2}}{1 - \sqrt{1-\gamma_3}\sqrt{1-\kappa_3}e^{-\frac{\alpha}{2}L_2 - jk_nL_2}} \tag{2.85}$$

where $L_2 = 2\pi R_L$ and R_L is the radius of the left side nano-ring. Regarding further simplification such as $x_1 = (1-\gamma_1)^{1/2}$, $x_2 = (1-\gamma_2)^{1/2}$, $y_1 = (1-\kappa_1)^{1/2}$, and $y_2 = (1-\kappa_2)^{1/2}$, the interior circulated light fields E_1, E_3 and E_4 are given by

$$E_1 = \frac{jx_1\sqrt{\kappa_1}E_{i1} + jx_1x_2y_1\sqrt{\kappa_2}E_{0L}E_{i2}e^{-\frac{\alpha}{2}\frac{L}{2} - jk_n\frac{L}{2}}}{1 - x_1x_2y_1y_2E_{0r}E_{0L}e^{-\frac{\alpha}{2}L - jk_nL}} \tag{2.86}$$

$$E_3 = x_2y_2E_{0r}E_1e^{-\frac{\alpha}{2}\frac{L}{2} - jk_n\frac{L}{2}} + jx_2\sqrt{\kappa_2}E_{i2} \tag{2.87}$$

$$E_4 = x_2y_2E_{0r}E_{0L}E_1e^{-\frac{\alpha}{2}L - jk_nL} + jx_2\sqrt{\kappa_2}E_{0L}E_{i2}e^{-\frac{\alpha}{2}\frac{L}{2} - jk_n\frac{L}{2}} \tag{2.88}$$

Thus, the throughput port (E_{t1}) output is expressed by

$$E_{t1} = AE_{i1} - BE_{i2}e^{-\frac{\alpha}{2}\frac{L}{2}-jk_n\frac{L}{2}}\left[\frac{CE_{i1}\left(e^{-\frac{\alpha}{2}\frac{L}{2}-jk_n\frac{L}{2}}\right)^2 + DE_{i2}\left(e^{-\frac{\alpha}{2}\frac{L}{2}-jk_n\frac{L}{2}}\right)^3}{1 - F\left(e^{-\frac{\alpha}{2}\frac{L}{2}-jk_n\frac{L}{2}}\right)^2}\right] \qquad (2.89)$$

where, $A = x_1x_2$, $B = x_1x_2y_2\sqrt{\kappa_1}E_{0L}$, $C = x_1^2x_2\kappa_1\sqrt{\kappa_2}E_{0r}E_{0L}$, $D = (x_1x_2)^2y_1y_2\sqrt{\kappa_1\kappa_2}$ $E_{0r}E_{0L}^2$ and $F = x_1x_2y_1y_2E_{0r}E_{0L}$. The power output of the throughput port (P_{t1}) is given by

$$P_{t1} = (E_{t1}) \cdot (E_{t1})^* = |E_{t1}|^2 \qquad (2.90)$$

Similarly, the output optical field of the drop port (E_{t2}) is given by Ali *et al.* (2010i) and Amiri *et al.* (2013b)

$$E_{t2} = x_2y_2E_{i2}\left[\frac{x_1x_2\sqrt{\kappa_1\kappa_2}E_{0r}E_{i1}e^{-\frac{\alpha}{2}\frac{L}{2}-jk_n\frac{L}{2}} + x_1x_2^2y_1y_2\sqrt{\kappa_2}E_{0r}E_{0L}E_{i2}\left(e^{-\frac{\alpha}{2}\frac{L}{2}-jk_n\frac{L}{2}}\right)^2}{1 - x_1x_2y_1y_2E_{0r}E_{0L}\left(e^{-\frac{\alpha}{2}\frac{L}{2}-jk_n\frac{L}{2}}\right)^2}\right]$$

$$(2.91)$$

where the power output of the drop port (P_{t2}) is expressed by

$$P_{t2} = (E_{t2}) \cdot (E_{t2})^* = |E_{t2}|^2 \qquad (2.92)$$

2.12 FIBER NONLINEARITIES

Fiber nonlinearity is the response of any dielectric to light under intense electromagnetic fields. The source of nonlinear reaction is related to harmonic motion of bound electrons under the influence of an applied field. The total polarization P induced by electric dipoles is not linear in the electric field (Butcher & Cotter, 1991).

$$\vec{P} = \varepsilon_0\left(\chi^1 \cdot \vec{E} + \chi^2 : \vec{E}\vec{E} + \chi^3 \vdots \vec{E}\vec{E}\vec{E} + \ldots\right), \qquad (2.93)$$

Here $\chi^{(j)}$ $(j = 1, 2, 3, \ldots)$ is j^{th} order susceptibility and ε_0 is the vacuum permittivity. In general, $\chi^{(j)}$ is a tensor of rank $j + 1$. The linear susceptibility $\chi^{(1)}$ represents the dominant contribution to the polarization. It affects the refractive index n and the attenuation coefficient α. The second-order susceptibility $\chi^{(2)}$ is responsible for nonlinear effects as second-harmonic generation. However, it is non zero only for media that lack of inversion symmetry at the molecular level. As SiO_2 is a symmetric molecule, $\chi^{(2)}$ vanishes for silica glasses. Optical fibers do not normally exhibit second-order nonlinear effects. The electric-quadrupole and magnetic-dipole moments generate weak second-order nonlinear effects. Defects or color centers inside the fiber core can also contribute to second-harmonic generation under certain conditions. The lowest-order nonlinear effects in optical fibers originate from the third-order susceptibility $\chi^{(3)}$, which is responsible for phenomena such as third-harmonic generation, four-wave mixing, and nonlinear refraction.

2.13 CALCULATION OF NONLINEAR REFRACTIVE INDEX

The optical Kerr effect causes a variation in index of refraction which is proportional to the local radiance of the light. This refractive index variation is responsible for the nonlinear optical effects of self-focusing, self-phase modulation and modulational and basis for Kerr-lens model locking. This effect is only significant with very intense beams such as lasers. In the optical Kerr effect, an intense beam of light in a medium provides the modulating electric field, without an external field. In this case, the electric field is given by:

$$E = E_\omega \cos(\omega t), \tag{2.94}$$

where E_ω is the amplitude of the wave. Combining this with the equation for the polarization, $\chi^3 |E_\omega|^3$:

$$P \cong \varepsilon_0 \left(\chi^1 + \frac{3}{4} \chi^3 |E_\omega|^2 \right) E_\omega \cos(\omega t). \tag{2.95}$$

This is in fact linear susceptibility with an additional non-linear term:

$$\chi = \chi_{\text{Lin}} + \chi_{\text{Nl}} = \chi^1 + \frac{3}{4} \chi^3 |E_\omega|^2, \tag{2.96}$$

and

$$n = (1 + \chi)^{1/2} = (1 + \chi_{\text{Lin}} + \chi_{\text{Nl}})^{1/2} \approx n_0 \left(1 + \frac{1}{2n_0^2} \chi_{\text{Nl}} \right) \tag{2.97}$$

where $n_0 = (1 + \chi_{\text{Lin}})^{1/2}$ is the linear refractive index. Using Taylor expansion $\chi_{\text{Nl}} \langle\langle n_0^2$, this gives an intensity dependent refractive index:

$$n = n_0 + \frac{3}{8n_0} \chi^3 |E_\omega|^2 = n_0 + n_2 I \tag{2.98}$$

Here n_2 is the nonlinear-index coefficient related to χ^3, and I is the intensity of the wave. The refractive index changes and is proportional to the intensity of the light travelling through the medium (Melnichuk & Wood, 2010). The tensorial nature of χ^3 can affect the polarization properties of optical beams through nonlinear bire-fringence. The values of n_2 are relatively small for most materials, on the order of $10^{-20} \, \text{m}^2 \, \text{W}^{-1}$ for typical glasses. Therefore beam intensities (irradiances) of the order of $1 \, \text{GW} \, \text{cm}^{-2}$ as produced by lasers generate significant variations in the refractive index via the Kerr effect. The optical Kerr effect manifests itself temporally as self-phase modulation, a self-induced phase- and frequency-shift in a pulse of light as it travels through a medium. This process, along with the dispersion, can produce opti-cal solitons. Spatially, an intense beam of light in a medium produces a change in the medium's refractive index that mimics the transverse intensity pattern of the beam. For example, a Gaussian beam result in a Gaussian refractive index profile, similar to that of a gradient-index lens. This causes the beam to focus itself, a phenomenon known as self-focusing which is responsible for phenomena such as third-harmonic generation, four-wave mixing, and nonlinear refraction (Dharmadhikari et al., 2009).

2.14   NONLINEAR SCHRODINGER EQUATION (NLS EQUATION)

The beam is assumed to propagate along the z axis and diffract (or self-focus) along the two transverse directions X and Y, where X, Y, and Z are the spatial coordinates associated with r. The function A (X, Y, Z) describes the evolution of the beam envelope. The intensity dependence of the refractive index effects considerably the propagation of electromagnetic waves. The field propagating in the z direction can be expressed as:

$$E(r,t) = A(r)\exp(i\beta_0 z),\tag{2.99}$$

where $\beta_0 = \kappa_0 n_0$ is the propagation constant in terms of the optical wavelength $\lambda = \frac{2\pi c}{\omega_0}$. In general it depends on z because fields change their shape while propagating. If the expression of the electric field has replaced in the Helmholtz Equation to solve, assuming that the envelope $A(r,t)$ changes slowly while propagating. When the nonlinear and diffractive effects are included and the envelope $A(r,t)$ is assumed to vary with z on a scale much longer than the wavelength λ. In this paraxial approximation the second derivative $\frac{\partial^2 A}{\partial z^2}$ can be neglected. Thus the beam envelope is found to satisfy the following nonlinear parabolic equation:

$$2i\beta_0\frac{\partial A}{\partial Z} + \left(\frac{\partial^2 A}{\partial X^2} + \frac{\partial^2 A}{\partial Y^2}\right) + 2\beta_0 k_0 n_{nl}(I)A = 0\tag{2.100}$$

In the absence of the nonlinear effects, this equation reduces to the well-known paraxial equation of scalar diffraction theory (Born et al., 1999). It is useful to introduce the scaled dimensionless variables as:

$$x = \frac{X}{w_0},\quad y = \frac{Y}{w_0},\quad z = \frac{Z}{L_d},\quad u = \sqrt{k_0|n_2|L_d}A\tag{2.101}$$

where w_0 is a transverse scaling parameter related to the input beam width and L_d is the diffraction length. In terms of these dimensionless variables, the form of a standard NLS equation can be presented as:

$$\frac{1}{2}\left(\frac{\partial^2 u}{\partial x^2} + \frac{\partial^2 u}{\partial y^2}\right) + i\frac{\partial u}{\partial z} \pm |u^2|u = 0,\tag{2.102}$$

The dimensionality of the NLS equation depends on the nature of the nonlinear medium. For example, when a nonlinear medium is in the form of a planar waveguide, the optical field is confined in one of the transverse directions by the waveguide itself. In the absence of the nonlinear effects, the beam spreads only along the x direction.

A planar waveguide supports a finite number of modes depends on its size. In this case, therefore:

$$E(r,t) = A(X, Z)B(Y) \exp(i\beta_0 z), \tag{2.103}$$

where the function $B(Y)$ describes the waveguide-mode amplitude and β_0 is the corresponding propagation constant. The normalized NLS Equation without second-order derivative is given as:

$$\frac{1}{2}\frac{\partial^2 u}{\partial x^2} + i\frac{\partial u}{\partial z} \pm |u^2|u = 0, \tag{2.104}$$

This constitutes the simplest form of the NLS equation. This equation can be solved exactly using the inverse scattering method for both signs of the nonlinear term. The bright and dark spatial solitons correspond to the choice of + and − signs, respectively. As a simple example of bright spatial solitons, consider Equation (2.104) with the plus sign for the nonlinear term, assuming that the CW beam is propagating inside a self-focusing Kerr medium. For plus sign the equation can be expressed as:

$$\frac{1}{2}\frac{\partial^2 u}{\partial x^2} + i\frac{\partial u}{\partial z} + |u^2|u = 0. \tag{2.105}$$

Although the inverse scattering method is necessary to find all possible solutions of Equation (2.105), the solution corresponding to the fundamental soliton can be obtained by solving the NLS equation directly without using this technique. This approach is applicable even when the inverse scattering method cannot be used. The approach consists of assuming that a shape-preserving solution of the NLS equation exists and has the form

$$u(z,x) = V(x) \exp(i\varphi(z,x)), \tag{2.106}$$

where V is independent of z to represent a soliton that maintains its shape during propagation. The phase φ depends on both z and x. If Equation (2.106) is substituted into Equation (2.105) and the real and imaginary parts are separated, we obtain two equations for V and φ. The phase equation shows that φ should be $\varphi(z,x) = K_z + p_x$, where K and p are constants. Physically p is related to the angle that the soliton trajectory forms with the z axis. Choosing $p = 0$, $V(x)$ is found to satisfy

$$\frac{d^2 V}{dx^2} = 2V(K - V^2). \tag{2.107}$$

This nonlinear equation can be solved by multiplying it by $2\left(\frac{dV}{dx}\right)$ and integrating over x. The result is:

$$\left(\frac{dV}{dx}\right)^2 = (2V^2 K - V^4 + C), \tag{2.108}$$

where C is a constant of integration. Using the boundary condition that both V and $\left(\frac{dV}{dx}\right)$ vanish as $|x| \to \infty$, C is found to be 0. The conditions that $V = a$ and $\left(\frac{dV}{dx}\right) = 0$, at the soliton peak, which is assumed to occur at $x = 0$, define the constant $K = \frac{a^2}{2}$, and hence $\varphi = \frac{a^2 z}{2}$. Equation (2.108) is easily integrated to obtain $V(x) = a \operatorname{sech}(ax)$, where a is the soliton amplitude. We have thus found that the $(1+1)$ dimensional NLS equation (2.146) has the following simple shape preserving solution:

$$u(z, x) = a \operatorname{sech}(ax) \exp(i\varphi(z, x)). \tag{2.109}$$

It represents the fundamental mode of the optical waveguide induced by the propagating beam. If the input beam has the correct shape, all of its energy will be contained in this mode, and the beam will propagate without change of its shape. If the input beam shape does not exactly match the shape, some energy will be coupled into higher-order bound modes of the nonlinear waveguide.

2.15 TEMPORAL SOLITON

An electric field is propagating in a medium shows the optical Kerr effect through a guiding structure that limits the power on the x–y plane. If the field propagates towards z with a phase constant β_0, then it can be expressed by Equation (2.110).

$$E(r, t) = Aa(t, z)f(x, y) \exp[i(\beta_0 z - \omega_0 t)] \tag{2.110}$$

with A as the amplitude of the field, $a(t, z)$ is the envelope that shapes the impulse in the time domain and $f(x, y)$ represents the shape of the field on the x–y plane. The Fourier transform of the electric field is:

$$\tilde{E}(r, \omega - \omega_0) = \int_{-\infty}^{\infty} E(r, t) \exp[-i(\omega - \omega_0)] t \, dt. \tag{2.111}$$

The complete expression of the field in the frequency domain has shown in Equation 2.112.

$$\tilde{E} = A\tilde{a}(\omega - \omega_0, z)f(x, y) \exp(i\beta_0 z). \tag{2.112}$$

The Helmholtz equation can be solved in the frequency domain in Equation (2.113).

$$\nabla^2 \tilde{E} + n^2(\omega)\kappa_0^2 \tilde{E} = 0 \tag{2.113}$$

The phase constant is expressed in Equation (2.114) as,

$$n(\omega)\kappa_0 = \beta(\omega) = \beta_0 + \beta_l(\omega) + \beta_{nl} = \beta_0 + \Delta\beta(\omega), \tag{2.114}$$

with a Taylor series centered on ω_0, the phase constant is expressed in Equation (2.115).

$$\beta(\omega) \approx \beta_0 + (\omega - \omega_0)\beta_1(\omega) + \frac{(\omega - \omega_0)^2}{2}\beta_2 + \beta_{nl} \tag{2.115}$$

where, as known:

$$\beta_m = \frac{d^m \beta(\omega)}{d\omega^m}\bigg|_{\omega=\omega_0} \tag{2.116}$$

If we assume the slowly varying envelope approximation

$$\left|\frac{\partial^2 \tilde{a}}{\partial z^2}\right| \langle\langle \left|\beta_0 \frac{\partial \tilde{a}}{\partial z}\right|. \tag{2.117}$$

The Equation (2.118) is achieved as:

$$2i\beta_0 \frac{\partial \tilde{a}}{\partial z} + \left[\beta^2(\omega) - \beta_0^2\right]\tilde{a} = 0. \tag{2.118}$$

For temporal soliton,

$$\beta^2(\omega) - \beta_0^2 \approx 2\beta_0 \Delta\beta(\omega). \tag{2.119}$$

Replacing Equation (2.119) in the Equation (2.118), the Equation (2.120) achieved easily,

$$i\frac{\partial \tilde{a}}{\partial z} + \Delta\beta(\omega)\tilde{a} = 0. \tag{2.120}$$

Expressing the products in term of the derivatives, Equation (2.121) can be achieved.

$$\Delta\beta(\omega) \Leftrightarrow i\beta_1 \frac{\partial}{\partial t} - \frac{\beta_2}{2}\frac{\partial^2}{\partial t^2} + \beta_{nl}. \tag{2.121}$$

The nonlinear component in terms of the amplitude of the field can be written as:

$$\beta_{nl} = K_0 n_2 I = K_0 n_2 n \frac{|A|^2}{2\eta_0}|a|^2 \tag{2.122}$$

For duality with the spatial soliton, they can be defined as,

$$L_{nl} = \frac{2\eta_0}{K_0 n n_2 |A|^2}. \tag{2.123}$$

The equation becomes:

$$i\frac{\partial a}{\partial z} + i\beta_1 \frac{\partial a}{\partial t} - \frac{\beta_2}{2}\frac{\partial^2 a}{\partial t^2} + \frac{1}{L_{nl}}|a|^2 a = 0. \tag{2.124}$$

This is actually a propagating among the z axis with a group velocity given by $v_g = \frac{1}{\beta_1}$, and provides information on the pulse changes in its shape while propagating. Substituting $T = t - \beta_1 z$ in Equation (2.124), gives:

$$i\frac{\partial a}{\partial z} - \frac{\beta_2}{2}\frac{\partial^2 a}{\partial T^2} + \frac{1}{L_{nl}}|a|^2 a = 0. \tag{2.125}$$

Solving Equation (2.124), bright soliton can be described by Equation (2.126).

$$a\left(\frac{T}{T_0}, \frac{z}{L_d}\right) = A\,\mathrm{sech}\left(\frac{T}{T_0}\right)\exp\left[i\left(\frac{z}{2L_d} - \omega_0 t\right)\right], \tag{2.126}$$

where A and z are the optical field amplitude and propagation distance, respectively. $T = t - \beta_1 \times z$, is a soliton pulse propagation time in a frame moving at the group velocity. The pulse is propagating between the z axis group velocity with $\left(v_g = \frac{1}{\beta_1}\right)$. $L_d = T_0^2/|\beta_2|$ is the dispersion length of the soliton pulse, where β_1 and β_2 are the coefficients of the linear and nonlinear dispersive as a second order and third order terms of the Taylor's expansion of the propagation constant. T_0 is a soliton pulse propagation time of the input signal. The t is the soliton phase shift and ω_0 is frequency shift of the soliton pulse. $I_{\max} = \frac{|\beta_2|}{T_0^2 \gamma}$ is the term of maximum optical soliton intensity.

For a soliton pulse, the dispersion length (L_d) and nonlinear length $L_{NL} = \left(\frac{1}{\gamma \phi_{NL}}\right)$ balances each other (Ali et al., 2010s; Amiri & Ali, 2013a; Mirzaee & Amiri, 2014; Amiri & Ali, 2014c; Akanbi et al., 2015).

2.16 GAUSSIAN BEAM

Gaussian pulse is a beam of electromagnetic wave that transfers electric field and intensity. Many lasers emit beams that approximates Gaussian shape, in which case the laser optical resonator. For a Gaussian pulse, the complex electric field amplitude is given by Equation (2.127).

$$E(r, z) = E_0 \frac{\omega_0}{\omega(z)}\exp\left(\frac{-r^2}{\omega^2(z)}\right)\exp\left(-ikz - ik\frac{r^2}{2R(z)} + i\xi(z)\right) \tag{2.127}$$

Here, r is the radial distance from the center axis of the beam, z is the axial distance from the beam's narrowest point. k is the wave number in a vacuum, $\omega(z)$ and E_0 is the radius at which the field amplitude and intensity drop to $1/e^2$ and $1/e$ of their axial values, respectively, ω_0 is the waist size, $R(z)$ is the radius of curvature of the beam and $\xi(z)$ is the phase shift, an extra contribution to the phase that is seen in Gaussian beams (Saleh et al., 1991). In the next section some parameters will be introduced to show the concept in this research.

2.17 DISPERSION

In optics, dispersion is the phenomenon in which the phase velocity of a wave depends on its frequency or alternatively when the group velocity depends on the frequency.

Dispersion represents a broad class of phenomena related to the fact that the velocity of the electromagnetic wave depends on the wavelength. In telecommunication the term of dispersion is used to describe the processes which cause that the signal carried by the electromagnetic wave and propagating in an optical fiber is degradated as a result of the dispersion phenomena. This degradation occurs because the different components of radiation having different frequencies propagate with different velocities. The waveguide dispersion is caused by the fact that some light travels in the fiber refractive index rather than the fiber core. Waveguide dispersion is also a type of chromatic dispersion (Mogilevtsev *et al.*, 1998).

2.18   GROUP VELOCITY DISPERSION

Group velocity dispersion is the phenomenon that the group velocity of light in a transparent medium depends on the optical frequency or wavelength. The term can also be used as a precisely defined quantity, namely the derivative of the inverse group velocity with respect to the angular frequency. In the theoretical treatment of pulse propagation (Boyd, 1992), it is often convenient to expand the propagation constant $\beta(\omega)$ in a power series about the central frequency ω_0 of the optical pulse as

$$\beta(\omega) = \beta_0 + \beta_1(\omega - \omega_0) + \frac{1}{2}\beta_2(\omega - \omega_0)^2 + \cdots \tag{2.128}$$

where $\beta(\omega_0) = \beta_0$ is the mean wave vector magnitude of the optical pulse,

$$\beta_1 = \left.\frac{d\beta}{d\omega}\right|_{\omega=\omega_0} = \frac{1}{v_g} \tag{2.129}$$

is the inverse of the group velocity, and

$$\beta_2 = \left.\frac{d^2\beta}{d\omega^2}\right|_{\omega=\omega_0} = \frac{1}{c}\frac{dn_g}{d\omega} \tag{2.130}$$

is a measure of the dispersion in the group velocity. Since the transit time through a material medium of length L is given by $T = \frac{L}{v_g} = L\beta_1$, the spread in transit time is given approximately by

$$\Delta T \cong L\beta_2\Delta\omega, \tag{2.131}$$

where $\Delta\omega$ is a measure of the frequency bandwidth of the pulse. The significance of each of the terms of the power series can be easily understood for example, by considering solutions to the wave equation for a transform-limited Gaussian-shaped pulse of characteristic pulse width T_0 incident upon a dispersive medium. When the propagation distance through the medium is much shorter than the dispersion length the pulse remains essentially undistorted and travels at the group velocity.

$$L_D = \frac{T_0^2}{|\beta_2|} \tag{2.132}$$

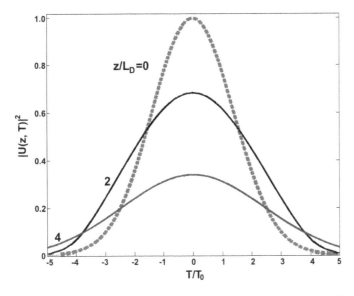

Figure 2.10 Effects of group velocity dispersion and higher-order dispersion on a Gaussian shaped pulse, the dashed curve shows the incident pulse envelope.

L_D is dispersion length. The dispersion length (L_D) and nonlinear length (L_{NL}) provides the length scale over dispersive or nonlinear effects which are important for pulse evolution. When fiber length L is such that $L \ll L_{NL}$ and $L \ll L_D$, neither dispersive nor nonlinear effects play a significant role during pulse propagation. The fiber plays a passive role in this regime and acts as a mere transporter for optical pulses. This regime is useful for optical communication systems. However, L_D and L_{NL} become smaller as pulses become shorter and more intense. When the fiber length L is longer or comparable to both L_D and L_{NL}, the dispersion and nonlinearity appear together as the pulse propagates along the fiber.

The interplay of the GVD and SPM effects lead to a qualitatively different behavior compared with that expected from GVD or SPM alone. In the anomalous-dispersion regime $(\beta_2 < 0)$, the fiber can support solitons. In the normal-dispersion regime $(\beta_2 > 0)$, the GVD and SPM effects can be used for pulse compression. For long propagation distances, that is shorter T_0 and larger $\Delta\omega$, the pulse broadens but retains its Gaussian shape as shown in Figure 2.10. In addition, the pulse acquires a linear frequency chirp, that is, the instantaneous frequency of the light varies linearly across the pulse about the central carrier frequency. T_0 is pulse propagation through a dispersive medium without significant pulse distortion, it is necessary that the spread of transit time's ΔT given by Equation (2.131) be much smaller than the characteristic pulse duration T_0.

2.19 SELF PHASE MODULATION (SPM)

SPM is a nonlinear optical effect of light-matter interaction. When a light intensity travels in a medium, it will induce a varying refractive index of the medium due to the optical Kerr effect. This variation in refractive index will produce a phase shift in the pulse, leading to a change of the pulse's frequency spectrum. SPM is an important optical phenomenon that uses short pulses of intense light, such as laser and optical fiber communication systems (Yan, 2010).

The third order susceptibility $\chi^{(3)}$ is the lowest-order nonlinear effects in optical fibers, responsible for phenomena such as third harmonic generation, four-wave mixing, and nonlinear refraction. The nonlinear processes involve the generation of new frequencies such as third-harmonic generation and four-wave mixing are not efficient in optical fibers. Most of the nonlinear effects in optical fibers originate from nonlinear refraction, a phenomenon referring to the intensity dependence of the refractive index. The refractive index has been achieved as mentioned in Equation (2.98), where n_0 is the linear part, I is the optical intensity inside the fiber, and n_2 is the nonlinear refractive index related to $\chi^{(3)}$.

The intensity dependence of the refractive index leads to a large number of interesting nonlinear effects. Self-phase modulation refers to the self-induced phase shift experienced by an optical field during its propagation in optical fibers. Phase of an optical field changes by

$$\phi = k_0 nL = k_0 n_0 L + k_0 n_2 IL = \phi_L + \phi_{NL} \tag{2.133}$$

where k_0 and L are wave number and fiber length. The intensity dependent nonlinear phase shift $(n_2 k_0 LI)$ is due to SPM. Among other things, SPM is responsible for spectral broadening of ultra-short pulses and formation of optical solitons in the anomalous-dispersion regime of fibers be much smaller than the characteristic pulse duration T_0 (Stolen & Lin, 1978).

2.20 CHAOTIC PHENOMENA

Chaotic signal is nonlinear property in physics, electronics and communication, which gives irregular behavior. Chaotic system provides a powerful mechanism for the design of be employed in various areas such as secured communication (Cuomo & Oppenheim, 1993). The chaotic communication has recently attracted great attention because of its potential application in communication security, where it uses a noise-like broadband waveform as a carrier (Ali et al., 2010d; Ridha et al., 2010b).

2.21 KRAMERS-KRONIG RELATIONS

The Kramers–Kronig relations are bidirectional mathematical relations, connecting the real and imaginary parts of any complex function that is analytic in the upper half-plane. These relations are often used to calculate the real part from the imaginary part (or vice versa) in physical systems (Toll, 1956). The Kramers–Kronig relations are derived by considering the integral as given in Equation (2.134).

$$I = P \int_{-\infty}^{+\infty} \frac{\chi(\omega')}{\omega' - \omega} dx \tag{2.134}$$

P shows the principal value of the integral and I represents the integral values. Here, ω' represents the complex angular frequency and $\chi(\omega')$ shows the dielectric susceptibility. And these Kramers–Kronig mathematical relations are used to examine the dispersion of material. For generation fast and slow light, the depressive medium or waveguide is required and the dispersion of the waveguide is tested by this method.

2.22 SCATTERING MATRIX METHOD FOR RING RESONATOR

An optical ring resonator is a set of waveguides in which at least one is a closed loop coupled some sort of light input and output. When a beam of light passes through a waveguide as shown in Figure 2.3 part of light will be coupled into the optical ring resonator. One frequently chosen way of modelling the response of a single micro-ring is the use of a scattering matrix (Capmany & Muriel, 1990) as illustrated in Figure 2.3. In the scattering matrix model the micro-ring is modelled as one coupler, which couple a fraction κ over to the cross and direct path. The optical fields in the inputs and outputs of the ring are related as follows:

$$\begin{bmatrix} E_1 \\ E_{\text{out}} \end{bmatrix} = \begin{bmatrix} A & B \\ B & A \end{bmatrix} \begin{bmatrix} E_2 \\ E_{\text{in}} \end{bmatrix} \tag{2.135}$$

$$E_1 = AE_2 + BE_{\text{in}} \tag{2.136}$$

$$E_{\text{out}} = BE_2 + AE_{\text{in}} \tag{2.137}$$

The parameters such as A and B are the coupler coefficient in direct and cross path as follows:

$$A = \sqrt{1-\kappa}, \quad B = j\sqrt{\kappa} \tag{2.138}$$

In this method the waveguide is symmetric and coupling is lossless. Lossless coupling is when no light is transmitted all the way through input waveguide to its own output and all of the light is coupled into the ring waveguide. For lossless coupling to occur, the following equation must be satisfied:

$$|A|^2 + |B|^2 = 1 \tag{2.139}$$

where A is the transmission coefficient through the coupler and B is the cross transmission coupling referred to as the coupling coefficient. With the scattering matrix model, the influence of the loss parameter on the micro-ring response can be determined as well. Parameters such as filtering bandwidth, insertion loss, crosstalk, and channel separation can be determined in this way.

2.23 THEORY OF SLOW LIGHT

The normalized group velocity as a function of $\omega_0 \frac{\partial n(\omega)}{\partial \omega}$ is shown in Figure 2.11. In the spectral region of normal dispersion, where $\frac{\partial n(\omega)}{\partial \omega} > 0$, the group velocity can be

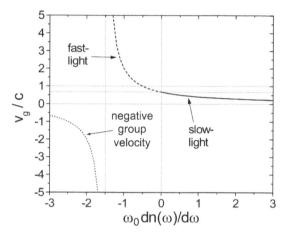

Figure 2.11 Solid bold lines represent slow-light, the dashed bold lines represent fast light and the dotted bold lines represent negative group velocities.

decreased. It is less than the phase velocity and can take on very low values. Since the pulse is decelerated, this corresponds to slow-light. If the refractive index slope is negative the group velocity is increased and observed anomalous dispersion. Therefore, the pulse will be accelerated and is faster than v_p. This is the region where fast light valid.

If $\omega_0 \frac{\partial n(\omega)}{\partial \omega}$ comes into the region of $n(\omega)$, $v_g \to \infty$ t_g becomes zero. The pulse delay equals the value of the phase delay, but with a negative sign. This means that the peak of the pulse arrives at the end of L at the same moment it enters the medium. For lower values of $\omega_0 \frac{\partial n(\omega)}{\partial \omega}$ the pulse travels with a negative group velocity and $t_g < 0$. Thus, the negative pulse delay further increases which means that a pulse travels backwards in the medium. In most cases, superluminal and negative group velocities are possible due to a pulse reshaping inside the medium which results in an advancement of the peak of the pulse. Such a reshaping can be based on an asymmetric absorption or amplification of the pulse energy for instance. Therefore, the pulse's peak is shifted if the trailing edge of the pulse experiences a higher absorption than the front edge or the trailing edge receives less amplification than the front edge (Wiatrek *et al.*, 2009). Although a superluminal and negative group velocity is very intriguing and extraordinary, such velocities do not violate Einstein's causality and special theory of relativity. As it will be explained in the next section, the group velocity is not the speed at which information propagates.

The alterable behavior of the group velocity can be used to engineer systems with large externally controllable dispersions, where $\frac{\partial n(\omega)}{\partial \omega}$ has very large positive or negative values. Thus, it is possible to propagate optical pulses extremely fast (Feng *et al.*, 2009) or extremely slow (Henker *et al.*, 2008a) and even to stop them completely (Junker *et al.*, 2007). These results with extreme values of v_g, especially the superluminal velocities, have revived the debates about the velocity of information. Slow and fast light effects invariably make use of the rapid variation of refractive index that occurs in the vicinity of a material resonance. Slow light can be achieved by making $\frac{\partial n(\omega)}{\partial \omega}$ large and positive values dispersion. The pulse is made up of several frequency components.

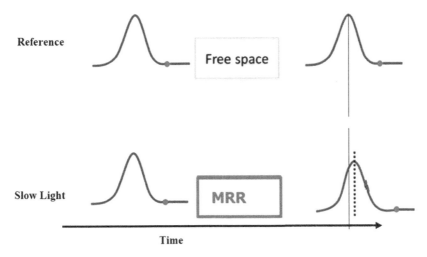

Figure 2.12 Propagation of a pulse in slow and fast light regimes.

It is the constructive interference of high frequency components that gives the pulse a sharp peak. If the pulse is travelling through a medium with $\frac{\partial n(\omega)}{\partial \omega} > 0$ that means the high frequency components are travelling slower than the low frequency components. Hence, the pulse peak appears to travel slowly through the medium (Bigelow *et al.*, 2003b).

In Figure 2.12, the leading edge of the pulse is indicated by the red dot. In all three cases, the leading edge of the pulse travels at the same velocity. However, in slow light the peak of the pulse is delayed while in the fast light the peak is advanced. Slow light is light that travels at an exceptionally slow group velocity, or in a medium with an exceptionally large group index. Traditionally, slow light is defined as having $V_g \ll V_p$, which occurs when $n_g \gg n_p$. Slow light techniques are aimed to increase n_g with increasing dispersion $\left(\frac{dn}{d\omega}\right)$. However, dispersion need not be positive. It can be zero, negative, or (nearly) infinite, leading to several other regimes of operation (Eliseev *et al.*, 2006).

2.24 OPTICAL BUFFER

An optical buffer is a device that is capable of temporarily storing light in telecommunication. Just as in the case of a regular buffer, it is a storage medium that enables compensation for a difference in time of occurrence of events. More specifically, an optical buffer serves to store data that is transmitted optically in the form of light without converting it into the electrical domain.

One is to investigate on the technological implementation of this buffer, and try to reduce the size by using slow-light devices. To obtain an optical buffer, in general, one must vary the medium within which the optical signal travels by either increasing the path length or reducing the signal group velocity. Optical delay time or optical

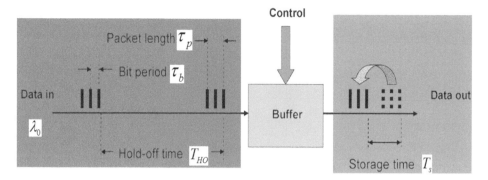

Figure 2.13 Concept of optical buffer memory.

buffers are the key components for future optical networks and information processing systems. In optical packet switch applications, buffers are required for synchronization of incoming packets and for collision avoidance on outgoing light paths.

At the heart of the buffer scaling problem is the issue of the physical size of a bit of data stored on an optical fiber delay line. Figure 2.13, shows some input and output optical data and defines the characteristics of the data and the buffer. The input data comprises a series of packets centered at the optical radian frequency ω_0. The free-space wavelength corresponding to ω_0 is λ_0. The length of the packets is τ_p, the bit period of the data in each packet is τ_b. The optical bandwidth of the data is $\Delta\upsilon = \Delta\omega/2\pi$. The number of bits in each packet is τ_p/τ_b. Two important parameters that characterize the overall performance of the buffer are the storage time (T_S) and the hold-off time (T_{HO}). In Fig. 3.8, the output packet shown with dashed lines represents the input data as it would have emerged from the buffer if it passed directly through the buffer without any controlled delay.

The storage time is the delay between this and the buffered packet. In Figure 2.13, the storage time T_S is larger than the packet length τ_p, but T_S can be smaller than τ_p. During the hold-off time, the buffer cannot accept any additional data. This limits the rate at which packets can be loaded into the buffer.

Chapter 3

Analysis of single Micro-Ring Resonators (MRR), add/drop filter MRR and cascaded MRR

3.1 SINGLE MICRO-RING RESONATOR (MRR)

Micro-Ring Resonators (MRRs) have great interest due to different applications especially in secured communication and light storage in optical buffers. Several methods have been reported for the generation of fast and slow light. In this work, micro-ring resonators are used to generate of slow and fast light. MRRs have been made from InGaAsP/InP, GaAlAs/GaAs and hydrogenated amorphous silicon (Aitchison *et al.*, 1992; Narayanan & Preble, 2010).

3.2 ANALYSIS OF SINGLE MICRO-RING RESONATOR (SMRR)

The schematic diagram of a nonlinear optics ring resonator system is shown in Fig. 3.2. The fiber coupler is connected to one ring of the resonator system. The nonlinearity of the fiber ring is of the Kerr-type. The refractive index of the proposed system is given by Ali *et al.* (2010e) and Shojaei & Amiri (2011a)

$$n = n_0 + n_2 I = n_0 + \left(\frac{n_2}{A_{\text{eff}}} \right) P, \tag{3.1}$$

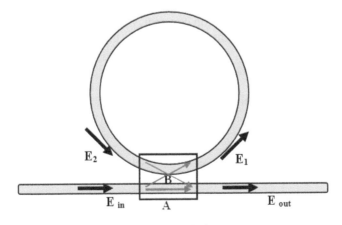

Figure 3.1 Scattering matrix model of a micro-ring resonator.

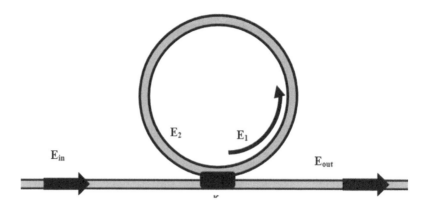

Figure 3.2 Schematic diagram of a single ring resonator with a single fiber coupler.

where n_2 is the nonlinear refractive index and n_0 is the linear refractive index. I and P are the optical intensity and optical field power, respectively. A_{eff} is the effective mode core area of the fiber. The study of nonlinear phenomena in fiber ring resonators plays an important role in optical communication. The nonlinearity leads to a decrease in the line width and absorption. Therefore these phenomena assist the generation of slow light in micro-ring resonator. For simplification of the equation, the output field at steady state is given as:

$$y_1 = \sqrt{1 - \kappa_1}, \quad x_1 = \sqrt{1 - \gamma_1}, \quad \tau = \exp(-\alpha L/2), \quad \phi = kLn_0 + kLn_2|E_1|^2 \quad (3.2)$$

Here, α and γ are the absorption coefficient and fractional coupling intensity loss respectively. L is the circumference of each ring and ϕ is the combination of linear and nonlinear phase shift. κ_1 is the coupling coefficient and $k = \frac{2\pi}{\lambda}$ is the wave propagation number in a vacuum (Zeinalinezhad *et al.*, 2014; Amiri & Ali, 2014d). τ is a one round trip loss. The transfer function of this configuration is derived by the scattering matrix method (Capmany & Muriel, 1990; Choi *et al.*, 2002). The light in the ring resonator filter is incorporated in the attenuation constant, the interaction can be described. The output electric field can be calculated via scattering matrix method as follows:

$$E_{\text{out}} = E_{\text{in}} y_1 x_1 + j\sqrt{\kappa_1} x_1 E_2 \tag{3.3}$$

E_1 and E_2 are in the electric field of ring resonators that are defined as

$$E_1 = j\sqrt{\kappa_1} x_1 E_{\text{in}} + E_2 x_1 y_1 \tag{3.4}$$

$$E_2 = E_1 \tau \exp(-j\phi) \tag{3.5}$$

Therefore,

$$E_1 = j\sqrt{\kappa_1} x_1 E_{\text{in}} + E_1 x_1 y_1 \tau \exp(-j\phi) \tag{3.6}$$

By solving this equation,

$$E_1 = \frac{j\sqrt{\kappa_1}x_1 E_{\text{in}}}{1 - x_1 y_1 \tau \exp(-j\phi)} \tag{3.7}$$

The output field can be written as:

$$E_{\text{out}} = E_{\text{in}} y_1 x_1 + j\sqrt{\kappa_1}x_1 \tau \exp(-j\phi)\left(\frac{j\sqrt{\kappa_1}x_1 E_{\text{in}}}{1 - x_1 y_1 \tau \exp(-j\phi)}\right) \tag{3.8}$$

From Equation (3.8), the transmission of ring can be written as:

$$T = \frac{E_{\text{out}}}{E_{\text{in}}} = \left(\frac{x_1 y_1 - x_1^2 \tau \exp(-j\phi)}{1 - x_1 y_1 \tau \exp(-j\phi)}\right) \tag{3.9}$$

T is the rate of transmission and the output power can be calculated as:

$$P_{\text{out}} \propto (E_{\text{out}}) \cdot (E_{\text{out}})^* = |E_{\text{out}}|^2 \tag{3.10}$$

The normalized transmission light field can be expressed (Ali *et al.*, 2010v; Amiri *et al.*, 2010; Amiri *et al.*, 2012a) as

$$\left|\frac{E_{\text{out}}}{E_{\text{in}}}\right|^2 = (1 - \gamma_1)\left[1 - \frac{\kappa_1\left[1 - (1-\gamma)^2\tau^2\right]}{1 + (1 - \gamma_1)^2(1 - \kappa_1)\tau - 2(1-\gamma_1)\sqrt{1-\kappa_1}\tau\cos\phi}\right]. \tag{3.11}$$

Equation (3.11) is a mathematical relation used for characterizing a nonlinear effects in micro-ring resonator such as bifurcation, chaos, and optical bistability.

3.3   SOLITON ROUNDTRIP AND ADD/DROP SYSTEM

The chaotic behavior of the multi output signals can be filtered using appropriate parameters of the add/drop filter system. Therefore the output signals are free from chaotic signals where multi soliton wavelength can be used to increase the capacity of the communication network link. For simplicity the parameters in obtaining the optical output at the throughput and drop port can be defined as follows (Ali *et al.*, 2010r; Amiri & Ali, 2014b):

$$y_1 = \sqrt{1 - \kappa_1}, \quad y_2 = \sqrt{1 - \kappa_2}, \quad x_1 = \sqrt{1 - \gamma_1}, \quad x_2 = \sqrt{1 - \gamma_2},$$
$$\tau = \exp(-\alpha L/2), \quad \phi = \phi_0 + \phi_{NL} \tag{3.12}$$

Here E_t and E_d represent the optical fields of the throughput and the drop port that they have shown in Figure 3.3. The transfer function of this configuration is derived by the scattering matrix method equation. The output electric field can be obtained as follows:

$$E_t = E_1 y_1 x_1 + j\sqrt{\kappa_1}x_1 E_b \exp(-\alpha L/4)\exp(-j\phi/2) \tag{3.13}$$

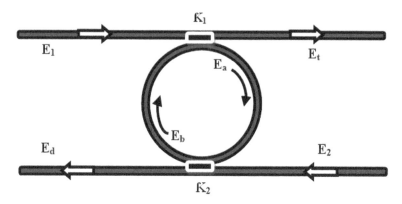

Figure 3.3 Schematic diagram for a ring resonator coupled to two waveguides, in an add/drop filter configuration.

$$E_d = j\sqrt{\kappa_2}x_2 E_a \exp(-\alpha L/4)\exp(-j\phi/2) + E_2 x_2 y_2 \tag{3.14}$$

The light fields in the add/drop filter system are given as:

$$E_a = j\sqrt{\kappa_1}x_1 E_1 + E_b x_1 y_1 \exp(-\alpha L/4)\exp(-j\phi/2) \tag{3.15}$$

$$E_b = E_a x_2 y_2 \exp(-\alpha L/4)\exp(-j\phi/2) + j\sqrt{\kappa_2}x_2 E_2 \tag{3.16}$$

From Equation (3.15), Equation (3.17)

$$E_a = \frac{j\sqrt{\kappa_1}x_1 E_1 + j\sqrt{\kappa_2}x_2 x_1 y_1 \exp(-\alpha L/4)\exp(-j\phi/2)E_2}{1 - x_1 y_1 x_2 y_2 \exp(-\alpha L/4)\exp(-j\phi/2)} \tag{3.17}$$

From Equation (3.16) in the Equation (3.18)

$$E_b = \frac{j\sqrt{\kappa_1}x_1 x_2 y_2 E_1 \exp(-\alpha L/4)\exp(-j\phi/2) + j\sqrt{\kappa_2}x_2 E_2}{1 - x_1 y_1 x_2 y_2 \exp(-\alpha L/2)\exp(-j\phi)} \tag{3.18}$$

Optical fields of the throughput and the drop port can be derived as:

$$E_d = jE_a \exp(-\alpha L/4)\exp(-j\phi/2)\sqrt{\kappa_2}x_2 + x_2 y_2 E_2 \tag{3.19}$$

Substituting Equation (3.17) in the Equation (3.19)

$$E_d = \left(\frac{\begin{array}{c} E_2 x_2 y_2 - \sqrt{\kappa_1 \kappa_2}x_1 x_2 E_1 \exp(-\alpha L/4)\exp(-j\phi/2) \\ - E_2 x_1 y_1 x_2^2 \exp(-\alpha L/2)\exp(-j\phi) \end{array}}{1 - x_1 y_1 x_2 y_2 \exp(-\alpha L/2)\exp(-j\phi)} \right) \tag{3.20}$$

The equation for the throughput result is presented by (3.21) achieved (Saktioto *et al.*, 2010b; Afroozeh *et al.*, 2014; Amiri *et al.*, 2014f),

$$
E_t = \left(\frac{\begin{array}{l} E_1 x_1 y_1 - E_1 x_1^2 x_2 y_2 \exp(-\alpha L/2) \exp(-j\phi) \\ - E_2 \sqrt{\kappa_1 \kappa_2} x_1 x_2 \exp(-\alpha L/4) \exp(-j\phi/2) \end{array}}{1 - x_1 y_1 x_2 y_2 \exp(-\alpha L/2) \exp(-j\phi)} \right) \tag{3.21}
$$

Output power for throughput and the drop port of the ring are:

$$
P_d \propto \left(\frac{\begin{array}{l} E_1^2[\kappa_1 \kappa_2 x_1^2 x_2^2 \exp(-\alpha L/2)] + E_2^2[x_2^2 y_2^2 - 2x_1 y_1^2 x_2^2 \\ \exp(-\alpha L/2) \cos\phi + x_1^2 y_1^2 x_2^2 \exp(-\alpha L)] + 2E_1 E_2 \sqrt{\kappa_1 \kappa_2} \\ \exp(-\alpha L/4) \cos(\phi/2)[x_1^2 y_1 x_2^3 \exp(-\alpha L/2) - x_1 x_2^2 y_2] \end{array}}{1 - 2x_1 y_1 x_2 y_2 \exp(-\alpha L/2) \cos(\phi) + x_1^2 y_1^2 x_2^2 y_2^2 \exp(-\alpha L)} \right). \tag{3.22}
$$

$$
P_t \propto \left(\frac{\begin{array}{l} E_2^2[\kappa_1 \kappa_2 x_1^2 x_2^2 \exp(-\alpha L/2)] + E_1^2[x_1^2 y_1^2 - 2x_1^3 y_1 y_2 x_2 \\ \exp(-\alpha L/2) \cos\phi + x_1^4 y_2^2 x_2^2 \exp(-\alpha L)] + 2E_1 E_2 \sqrt{\kappa_1 \kappa_2} \\ \exp(-\alpha L/4) \cos(\phi/2)[x_1^3 y_2 x_2^2 \exp(-\alpha L/2) - x_2 x_1^2 y_1] \end{array}}{1 - 2x_1 y_1 x_2 y_2 \exp(-\alpha L/2) \cos\phi + x_1^2 y_1^2 x_2^2 y_2^2 \exp(-\alpha L)} \right). \tag{3.23}
$$

Here, $\exp(-\alpha L/2)$ is a roundtrip loss coefficient. The circumference of each ring is $L = 2\pi R$ and R is the radius of the ring. $\phi = \beta L$ is the phase constant. κ_1 and κ_2 are the coupling coefficients of the add drop filter system as shown in Figure 3.3. The results from this system show a series of filtered multi wavelengths with defined FSR and FWHM characters. The important parameters of the system such as coupling coefficients, radius of the ring and center wavelength of the input pulse can vary in such a way the FSR can be increased and the FWHM is expected to be decreased for enhancing capacity and secured communication.

3.4 CHARACTERISTICS OF THE RING RESONATOR

There are a number of important characteristics of the ring resonator which include the FSR (Free Spectral Range), FWHM (Full Width at Half Maximum) and Finesse and Quality factor (Q factor). In the following section, these quantities are defined.

3.5 FREE SPECTRAL RANGE (FSR)

FSR is the separation of successive channels. The separation between two consecutive resonant peaks at the drop port is known as FSR. It is the spacing in optical frequency or wavelength between two successive reflected or transmitted optical intensity maxima or minima of an interferometer or the diffractive optical element as shown in Fig. 3.6.

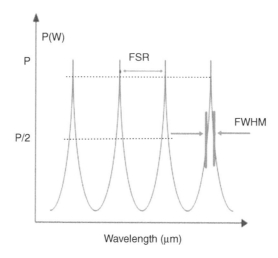

Figure 3.4 Schematic of FSR and FWHM.

At resonance, $\omega T_R = 2\pi M$, where T_R is the round-trip time, and M is an integer. The two successive resonances, ω_1 and ω_2, are related to each other as described by Kim *et al.* (2007).

$$\text{FSR} = \frac{\lambda_0^2}{Ln_g} \tag{3.24}$$

3.6 FULL WIDTH AT HALF MAXIMUM (FWHM)

The FWHM is an expression of the extent of a function, given by the difference between the two extreme values of the independent variable at which the dependent variable is equal to half of its maximum value as shown in Figure 3.4. The FWHM is used for such phenomena as the duration of pulse waveforms and the spectral width of sources used for optical communication and the resolution of spectrometers.

3.7 FINESSE

The finesse of an optical resonator is defined as its FSR range divided by the FWHM bandwidth of its resonances. It is fully determined by the resonator losses and is independent of the resonator length. The finesse gives the resolving power of the resonator when used as a transmission filter. An interesting fact is that a resonator finesse is independent of its dimension or circulating light wavelength (Chang & Sirkis, 1996) and described as:

$$\text{Finesse} = \frac{\text{FSR}}{\text{FWHM}} \tag{3.25}$$

3.8   QUALITY FACTOR (Q FACTOR)

The quality factor is a dimensionless parameter that describes how under damped an oscillator or resonator is, or equivalently, characterizes a resonator's bandwidth relative to its center frequency.

$$Q = \frac{\omega}{\Delta\omega} \tag{3.26}$$

3.9   GROUP VELOCITY AND PHASE VELOCITY

By considering a monochromatic plane wave of angular frequency ω propagating in a medium of refractive index n, the wave can be described by

$$E(z,t) = A \exp(i(kz - \omega t)), \tag{3.27}$$

where $k = \frac{n\omega}{c}$. We define the phase velocity v_p to be the velocity at which points of constant phase move through the medium. Since the phase of this wave is clearly given by

$$\varphi = (kz - \omega t), \tag{3.28}$$

Points of constant phase move a distance Δz in a time Δt, which are related by

$$k\Delta z = \omega\Delta t. \tag{3.29}$$

Thus:

$$v_p = \frac{\Delta z}{\Delta t} = \frac{\omega}{k} = \frac{c}{n}. \tag{3.30}$$

Here, the propagation of a pulse through a material system is considered. A pulse is composed of a spread of optical frequencies, at the peak of the pulse, the various Fourier components will tend to add up in phase. If this pulse is to propagate without distortion, these components must add in phase for all values of the propagation distance z. To express this thought mathematically, we first write the phase of the wave as:

$$\varphi = \frac{n\omega z}{c} - \omega t, \tag{3.31}$$

And require that there is no change in φ to first order in ω, that is:

$$\frac{d\varphi}{d\omega} = 0, \tag{3.32}$$

or

$$\frac{dn}{d\omega}\frac{\omega z}{c} + \frac{nz}{c} - t = 0, \tag{3.33}$$

which can be written as $z = v_g t$ where the group velocity is given by

$$v_g = \left(\frac{\partial \omega}{\partial k}\right) = \frac{c}{n(\omega) + \omega \frac{\partial n}{\partial \omega}}. \tag{3.34}$$

Here $n(\omega)$ is the refractive index and n_g is called the group index. The wave number k can be considered as the change in spectral phase per unit length. The group velocity is the velocity in which the envelope of a pulse propagates in a medium, assuming a long pulse with narrow bandwidth and the absence of nonlinear effects. The last equality in this equation results from the use of the relation $k = \frac{n\omega}{c}$. Alternatively, in terms of a group refraction index n_g can be defined by

$$v_g = \frac{c}{n_g} \tag{3.35}$$

$$n_g = n(\omega) + \omega \frac{\partial n}{\partial \omega} \tag{3.36}$$

Note that the group index differs from the phase index by a term that depends on the dispersion $dn/d\omega$ of the refractive index.

3.10 SEMICONDUCTOR CASCADED MRR ANALYSIS AND CHARACTERIZATION

3.10.1 Introduction of optical filters MRRs

The Micro-Ring Resonator (MRR) use light and follow the principles behind constructive interference and total internal reflection (Amiri et al., 2012b; Afroozeh et al., 2014; Amiri & Afroozeh, 2014). When light of the resonant wavelength is passed through the system from the input port, it builds up in intensity over multiple round-trips due to constructive interference and is output to the output port which serves as a detector waveguide (Amiri et al., 2014e; Amiri et al., 2014f; Amiri & Ahmad, 2014). The optical MRR functions as a filter because only a select few wavelengths will be at resonance within the loop. The MRR can be integrated with two or more ring waveguides to form an add-drop filter system or cascaded MRRs (Amiri et al., 2014b; Alavi et al., 2014; Amiri et al., 2015a). Optical filters are designed based on electromagnetic models to solve the fields in the frequency/wavelength or time domain (Alavi et al., 2014a; Amiri et al., 2015a; Amiri et al., 2015b). Optical filter is generally acted as an interferometer which cleaves the input signal into several paths with delaying, recombining and wavelength independent approach. The variation in splitting and recombining ratios and delays leads to change in the frequency response. These filters are mostly characterized by their frequency response (Dey & Mandal, 2012). In the context of signal processing, several analytical methods, such as the scattering matrix method (Schwelb, 1998; Madsen & Zhao, 1999) and the transfer-matrix-chain-matrix algebraic method (Moslehi et al., 1984; Capmany & Muriel, 1990) have been introduced to determine optical filter transfer functions in the Z-domain. The Vernier operation with Signal Flow Graph (SFG) method is a graphical approach for analyzing the intricate photonic circuits and quick calculation of optical transfer function. The SFG technique has some distinct advantages such as the graphical representation of signals behavior with the optical system. It is able to provide simple and a systematic technique of controlling the system's variables.

3.11 THEORETICAL BACKGROUND OF CASCADED MRR SYSTEM

The mathematics solution of the MRR system is based on the Vernier effect calculations for the CMRR. A resonating layout including a double stage MRR with 2×2 optical couplers which are vertically coupled together is shown in Figure 3.1(a). The signal flow graph (SFG) diagram of 2×2 optical directional couplers is displayed in Figure 3.1(b). By taking into account the insertion loss γ and the coupling factor k_i of the ith coupler ($i = 1, 2, 3$ for each coupler), the fraction of light pass through the throughput path is expressed as $C_i = \sqrt{(1 - \gamma_i)(1 - k_i)}$ and in contrast, the fraction of light pass through the cross path is expressed as $S_i = \sqrt{(1 - \gamma_i) k_i}$ (Mandal *et al.*, 2006; Bahadoran *et al.*, 2013a). The Z-transform parameter is defined as $Z^{-1} = \exp(-j 2\pi n_{\text{eff}} L/\lambda)$, where n_{eff} is the effective refractive index of the waveguide (Alavi *et al.*, 2015b), λ is the center wavelength and the circumference of the ring is $L = 2\pi R$, here R represents the radius of the MRR (Amiri *et al.*, 2013a; Afroozeh *et al.*, 2015; Amiri *et al.*, 2015c). Based on the Mason's rule the optical transfer function, H, for an optical device with the input photonics node $E_i(z)$ and the output photonics node $E_n(z)$ is

$$H = \frac{E_n(z)}{E_i(z)} = \frac{1}{\Delta} \sum_{j=1}^{n} T_j \, \Delta_j \tag{3.37}$$

where T_j shows the gain of the ith forward path from the input to output port and n is the overall number of onward paths from input to output photonics nodes. The symbol Δ_j considers all of the loops that remain untouched while a signal transverses via each T_j forward path from input to output photonics nodes. The signal flow graph determinant is displayed by Δ, which is given by (Dey *et al.*, 2013).

$$\Delta = 1 - \sum_{i=1} L_i + \sum_{i \neq j} L_i L_j - \sum_{i \neq j \neq k} L_i L_j L_k + \cdots \tag{3.38}$$

Here L_i is the transmittance gain of the ith loop. The SFG for our proposed system is illustrated in Figure 3.5(b), in which the input node is $E_1 = E_{\text{in}}$ and $E_{12} = E_{\text{drop}}$ are considered as the drop node. The Free Spectral Range (FSR) of the device is determined by FSR $= c/n_g L$ where $n_g = n_{\text{eff}} + f_0 (dn_{\text{eff}}/df)_{f_0}$ is the group refractive index of the ring, n_{eff} is the effective refractive index, and f_0 is the design (center) frequency (Sirawat-tananon *et al.*, 2012; Amiri *et al.*, 2013e; Amiri & Ali, 2014c; Amiri & Ali, 2014c; Amiri *et al.*, 2014d). The FSR of the CMRR with different radii can be determined by (Rabus *et al.*, 2005; Bahadoran *et al.*, 2013b).

$$\text{FSR}_{\text{tot}} = N_1 \cdot \text{FSR}_1 = N_2 \cdot \text{FSR}_2 \tag{3.39}$$

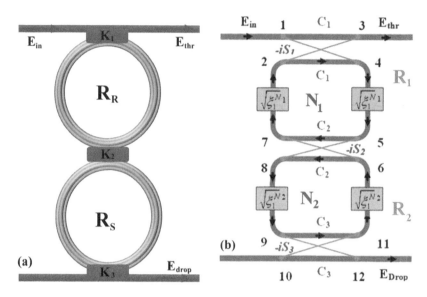

Figure 3.5 CMRR configuration (a) waveguide layout and (b) Z-transform diagram SFG.

where N_1 and N_2 are integer resonant mode numbers of each rings which can be determined by the ratio of FSR_{tot} rather than the FSR of each rings (FSR_i). To determine the Optical Transfer Function (OTF), non-touching loops and forward transmittance paths have to be identified from SFG diagram. The CMRR system is shown in Figure 3.5.

From Figure 3.5(b), three individual loops can be found for the CMRR as Equation (3.40). Two separate and non-touching transmittance loops from L_1 and L_2 exist as Equation (3.41). Three onward route transmittances with their delta determinant, Δ_i, recognizes from input node 1 to through node 3. The first transmittance onward path belongs to direct route which pass via (1→3) photonics nodes is given by Equation (3.42). The four non-touching loops also can be considered for this path so the delta determinant for this rout is given by Equation (3.43).

The second route goes to the track passes via these photonics nodes (1→ 4 → 5 → 7 → 2 →3). Hence, the second transmittance path is expressed by Equation (3.44). For this track only one non-touching loop, L_2, can be found which make the delta determinant as shown by Equation (3.45). The third transmittance route is the track which traverses via both rings through (1→ 4 → 5 → 8 → 9 → 11 → 6 → 7 → 2 →3) photonics nodes shown by Equation (3.46). There is not any non-touching loop for this track so the delta determinant can be expressed by Equation (3.47). Considering coupling loss $C_i^2 + S_i^2 = 1 - \gamma_i$ in addition to use the Mason rule for CMRR filter, the OTF for throughput port of CMRR Vernier filter can be calculated as by Equation (3.48). The same procedure can be done for drop port of the CMRR but transmittance forward paths changes to a track passes from (1→ 4 → 5 → 8 → 9 → 12) nodes.

The drop port transmittance path changes to Equation (3.49). Since this path touches all loops, the delta determinant for this path is equal to one shown by

Equation (3.50). Employing Mason's rule for drop port, the OTF for drop port CMRR Vernier filter is expressed by equation (3.51).

$$L_1 = C_1 C_2 \xi_1^{N1}, \quad L_2 = C_2 C_3 \xi_2^{N2}, \quad L_3 = -C_1 C_3 S_2^2 \xi_1^{N1} \xi_2^{N2} \tag{3.40}$$

$$L_4 = L_1.L_2 = C_1 C_2^2 C_3 \xi_1^{N1} \xi_2^{N2} \tag{3.41}$$

$$T_1^{thr} = C_1 \tag{3.42}$$

$$\Delta_1^{thr} = 1 - (L_1 + L_2 + L_3) + L_4 \tag{3.43}$$

$$T_2^{thr} = -S_1^2 C_2 \xi_1^{N1} \tag{3.44}$$

$$\Delta_2^{thr} = 1 - L_2 = 1 - C_2 C_3 \xi_2^{N2} \tag{3.45}$$

$$T_3^{thr} = S_1^2 S_2^2 C_3 \xi_1^{N1} \xi_2^{N2} \tag{3.46}$$

$$\Delta_3^{thr} = 1 \tag{3.47}$$

$$H_{31}^{thr} = \frac{\{C_1 - (1-\gamma_1)C_2\xi_1^{N1} - C_1 C_2 C_3\xi_2^{N2} + (1-\gamma_1)(1-\gamma_2)C_3\xi_1^{N1}\xi_2^{N2}\}}{\{1 - C_1 C_2\xi_1^{N1} - C_2 C_3\xi_2^{N2} + (1-\gamma_2)C_1 C_3\xi_1^{N1}\xi_2^{N2}\}} \tag{3.48}$$

$$T_1^{drp} = -iS_1 S_2 S_3 \sqrt{\xi_1^{N1}}\sqrt{\xi_2^{N2}} \tag{3.49}$$

$$\Delta_1^{drp} = 1 \tag{3.50}$$

$$H_{31}^{thr} = \frac{-iS_1 S_2 S_3 \sqrt{\xi_1^{N1}}\sqrt{\xi_2^{N2}}}{\{1 - C_1 C_2\xi_1^{N1} - C_2 C_3\xi_2^{N2} + (1-\gamma_2)C_1 C_3\xi_1^{N1}\xi_2^{N2}\}} \tag{3.51}$$

To obtain optimum coupling for higher transmission in drop port, we supposed that the input signal is totally coupled into the ring resonator and the transmission in through port is zero, $H_{82}^{thr} = 0$ (Amiri *et al.*, 2012c; Amiri *et al.*, 2013d; Amiri *et al.*, 2014g; Amiri & Afroozeh, 2014a). For further simplification with considering the exponential series up to the 1st order we suppose $\xi_1^{N1} = \xi_2^{N2} \approx 1$ as the imaginary part will vanish in resonance condition. For a choice of $k_2 = k_3$ the value of k_1 is determined as

$$k_1 = 1 - \frac{(1-\gamma_2)(1-\gamma_1)(1-k_2)[1-(1-\gamma_2)]^2}{[1-(1-k_2)(1-\gamma_2)]^2} \tag{3.52}$$

where γ_i represent the intensity insertion loss coefficients for couplers between rings and the bus waveguides (Amiri & Ali, 2013b; Amiri & Ali, 2014b; Amiri *et al.*, 2014a; Amiri & Afroozeh, 2014b; Alavi *et al.*, 2014b). In this chapter, we apply the Vernier operation with Signal Flow Graph (SFG) which is a graphical approach for analyzing the intricate photonic circuits mathematically and quick calculation of optical transfer function. Analysis of a Cascaded Micro-Ring Resonators (CMRR) made of InGaAsP/InP semiconductor is presented using the Signal Flow Graph (SFG) method which enables modelling the transfer function of the passive CMRR. These passive filters are mostly characterized by their frequency response. The theoretical calculations of the system are performed by the Vernier effects analysis. Two MRRs with radius of 100 μm which are vertically coupled together are used to generate resonant peaks. Here, the phase, dispersion and group delay of the generated signals are analyzed.

3.12 PHASE AND DISPERSION RESPONSES AND GROUP DELAY ANALYSIS OF THE CASCADED MRR SYSTEM

Two MRRs with radius of 100 μm which are vertically coupled together are used to generate resonant peaks as shown in Figure 2.5(a). The resonating system are fabricated from III/V semiconductors (InGaAsP/InP) on the basis of InP with a direct bandgap (Sadegh Amiri *et al.*, 2013; Amiri & Ali, 2014a; Amiri *et al.*, 2015b). The propagation loss is 0.1 dB/cm and the waveguide cores are 0.25 μm² (Amiri & Ali, 2013; Amiri *et al.*, 2013b; Amiri *et al.*, 2014e; Amiri *et al.*, 2015b). The resonant mode numbers for this optical system get equal values since the material and the length of both rings are the same. The waveguide's intensity attenuation coefficient is $\alpha = 0.1$ dB/cm (Amiri *et al.*, 2011a; Amiri *et al.*, 2014c; Alavi *et al.*, 2015a), intensity insertion loss coefficients for couplers between rings and the bus waveguides are $\gamma_1 = 0.0001$ and $\gamma_2 = \gamma_3 = 0.001$. An often-used component in micro-ring-based optical circuitry is the directional coupler-a twin waveguide structure used to couple a fraction of light from one waveguide to another (Afroozeh *et al.*, 2014a; Amiri & Afroozeh, 2014c; Amiri & Afroozeh, 2014d). Directional couplers are used to transfer light into and out of a MRR, and can be designed with a high degree of accuracy. The coupling coefficients are selected to $\kappa_1 = \kappa_2 = \kappa_3 = 0.02$. The input laser power versus the wavelength variations is shown in Figure 3.6.

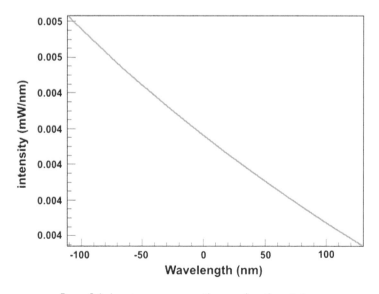

Figure 3.6 Input power versus the wavelength variation.

As the resonators are weakly coupled, an optical signal in the structure effectively takes a longer time to tunnel from resonator to resonator. We coupled light into the device by butt coupling a single-mode fiber to the facet. Figure 3.7 shows the through port phase response versus the wavelength and frequency respectively.

Figure 3.8 shows the drop port phase response versus the wavelength and frequency respectively.

Dispersion measures the rate of change of the group delay regarding to the wavelength (Amiri *et al.*, 2014c, 2014d, 2015c). Several factors contributors to dispersion. There is waveguide dispersion due to the fact that the electromagnetic wave is constrained to propagate in a guide of a given shape and cross sectional area. There is material dispersion due to the fact that the refractive indices involved are wavelength dependent. There is intermodal dispersion caused by the mixing of modes in a multi-mode system that is of no concern under single-mode operation. Finally there is structural dispersion that is determined by the architecture of the filter. The dispersion responses of the through port of the CMRR is shown in Figure 3.9.

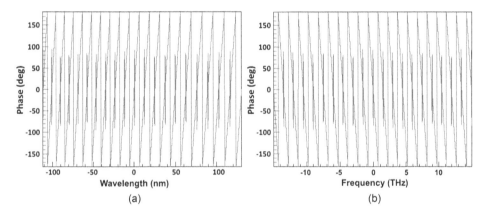

Figure 3.7 Through port phase response, (a): Phase response versus wavelength, (b): Phase response versus frequency.

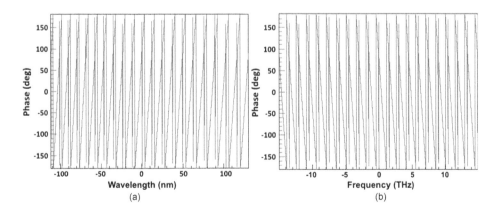

Figure 3.8 (a): Drop port phase response, (a): Phase response versus wavelength, (b): Phase response versus frequency.

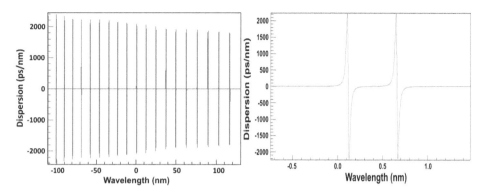

Figure 3.9 Through port dispersion response reference to the input port versus wavelength.

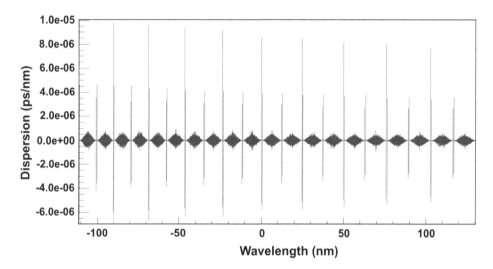

Figure 3.10 Through port dispersion response reference to the drop port versus wavelength.

The dispersion response of the through port reference to the drop port of the CMRR is shown in Figure 3.10.

Figure 3.11 shows the dispersion responses of the drop port reference to the input port of the CMRR.

Figure 3.12 shows the group delay of the drop port reference to the input port versus wavelength and frequency respectively, where Figure 3.13 shows the group delay reference to the through port versus wavelength and frequency respectively.

The performance of the passive ring resonators for filter application is limited by the internal losses. The incorporating of a semiconductor optical fiber (SOA) enables additional functionality such as the compensation of internal losses. Thus, the combination of a passive and active material enables the possibility to realize ring resonators

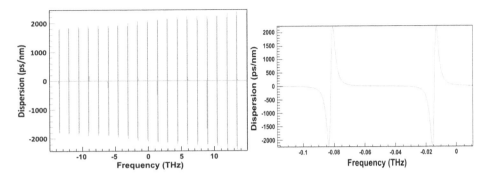

Figure 3.11   Drop port dispersion response reference to the input port versus frequency.

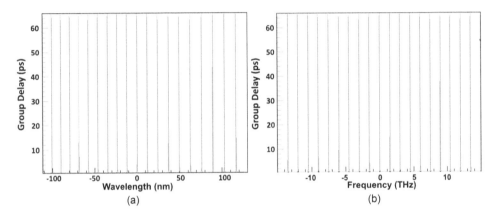

Figure 3.12   (a): Group delay (ps) of the drop port reference to the input port versus wavelength, (b): Group delay (ps) versus frequency.

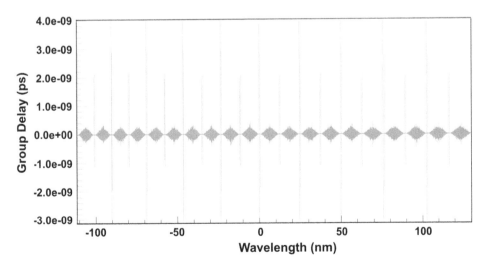

Figure 3.13   Group delay (ps) of the drop port reference to the through port versus wavelength.

with integrated SOA similar to the fiber optic filters with Erbium-Doped Fiber Amplifiers (EDFA), for improved filter performance of multi coupled ring resonator devices. Therefore, a Cascaded Micro-Ring Resonator (CMRR) is presented to show and analyze the phase, dispersion and group delay responses. This system consists of two MRRs vertically coupled which have the same radius. The input laser pulse is used to propagate within the MRR system. This system act as add/drop MRR system so that the spectrum of the input pulse will experiences the constructive and destructive interferences. The mathematics solution of the system is performed using the Vernier effect, where the Signal Flow Graph (SFG) is used to analyze the complex photonic circuits mathematically.

Chapter 4

Physics and fabrication of Micro-Ring Resonator (MRR) structure devices

4.1 INTRODUCTION

Optical micro-ring resonators have recently attracted increasing attention in the photonics community (Guarino *et al.*, 2007; Afroozeh, 2014c). Their applications range from quantum electro-dynamics to sensors and filtering devices for optical telecommunication systems, where they are likely to become an essential building block (Desurvire *et al.*, 2002; Afroozeh, 2014b). The integration of nonlinear and electro-optical properties in the resonators represents a very stimulating challenge, as it would incorporate new and more advanced functionality. Lithium niobate is an excellent candidate material, being an established choice for electro-optic and nonlinear optical applications (Holman *et al.*, 1987; Guarino *et al.*, 2007). Here we report on the first realization of optical micro-ring resonators in sub micrometric thin films of lithium niobate. The high index contrast films are produced by an improved crystal ion slicing and bonding technique using benzocyclobutene (Poberaj *et al.*, 2009; Afroozeh 2014). The rings have radius $R = 100\,\mu$m and their transmission spectrum has been tuned using the electro-optic effect. These results open integrated optical devices and nonlinear optical micro cavities (Tadigadapa & Mateti, 2009).

The established use of Wavelength Division Multiplexed (WDM) for local area network systems has raised the demand for new filtering and switching functions (Park *et al.*, 2004). In order to integrate these devices on a wafer scale, whispering gallery mode micro resonators represent the most compact and efficient solution. They consist of a bus waveguide evanescently coupled to a micrometer-size ring resonator; the characteristic size-dependent frequency spectrum of the ring allows only selected wavelength channels to be transmitted or shifted to another waveguide. Small radii allow a large free spectral range – i.e. large separation between the filtered channels but increase the propagation bending losses, which can compromise the quality factor (Melloni *et al.*, 2001). To overcome this limitation, high refractive index contrast between the ring core and the surrounding materials is mandatory (Knight *et al.*, 1998). A second, very important, requirement relates to the tenability. The possibility to electrically control the transmission spectrum, via electro-optic effect, would allow extremely compact and ultrafast modulation and switching (Scolari *et al.*, 2005). By integrating arrays of micro-ring resonators on a single optical chip, the realization of complex functions would be feasible. Besides, large-Q resonators based on non centro-symmetric materials would exploit the high amount of stored energy for enhancing the efficiency of nonlinear optical phenomena (Brochu & Pei, 2010).

Several examples of micro-ring resonators have been proposed and successfully realized in the last years in a variety of materials like semiconductors, silica and polymers (Little *et al.*, 1997a). The advanced structuring technology in semiconductor materials enables the realization of very high-Q resonators even for radii as small as 10 μm (Soltani *et al.*, 2007). Silicon-based resonators can be tuned by electrically-driven carriers injection in the core, but do not own truly nonlinear optical properties and their application is limited to infrared wavelengths (Soref, 2006; Afroozeh, 2014a). Polymers represent a very flexible solution in terms of processing and structuring, but the minimum resonator dimensions (and therefore the maximum achievable free spectral range) are limited by the low refractive index of the material. Silica rings, finally, do not provide any fast nonlinear or electro-optical property (Falcaro *et al.*, 2004; Afroozeh, 2014d).

A new technique, based on crystal ion slicing and wafer bonding, has been recently developed to produce sub-micrometric thin films of single-crystalline quality; it provides much higher refractive index contrast than the standard waveguide production methods in lithium niobate (Rabiei & Gunter, 2004). This is an essential asset for the fabrication of small radius ring resonators. An electro-optic modulator has been demonstrated by using lithium niobate films bonded to SiO2 as substrate (Rabiei & Gunter, 2004). However, the direct bonding method does not provide large area films and lacks of sufficient reproducibility, due to the severe requirements on the surface roughness and imperfections. Bonding of lithium niobate films to other substrates has also been reported, but suffers of film cracking due to the large mismatch between the thermal expansion coefficients of films and substrates and does not provide the optical contrast needed for the realization of optical micro-ring resonators (Takagi *et al.*, 1999).

The progress made in physics and technology of semiconductors depends mainly on two families of materials: the group IV elements and the III-V compounds. The first report of the formation of III-V compounds was published in 1910 by Thiel and Koelsch. They synthesized a compound of indium and phosphorus and reached the conclusion that its formula is very probably InP. As main application of semiconductor devices, the ring resonating systems can be fabricated from III/V semiconductors (InGaAsP/InP) on the basis of InP with a direct bandgap. A Rectangular Geometry Waveguide (RWG) is composed of one or more "slices" arranged side by side, with each slice being uniform in the horizontal direction but possibly possessing one or more layers in the vertical direction.

It is convenient for describing waveguides grown via an epitaxial method or more generally any waveguide that can be portrayed by a modest number of rectangles. The RWG also has a facility for simulating an etching process. A SWG (1D-slab waveguide structure) defines the vertical profile of a slice, and the same SWG can be used to describe all slices within a RWG representation. Accurate modeling results requires smooth waveguide sidewalls for low waveguide losses, deep etched curvatures for low bending losses, and precise waveguide dimensions for power splitting. In this case, lateral structure is achieved by varying the etch depth. The waveguide section is shown in Figure 4.1.

Slice Width (μm)	Slice etch Depth (μm)		Materials	Thickness	Mx:As-frac
2	0.9		InP	0.7	0
5	0.1		InGaAsP	0.8	0.05
2	0.9		InP	1.5	0

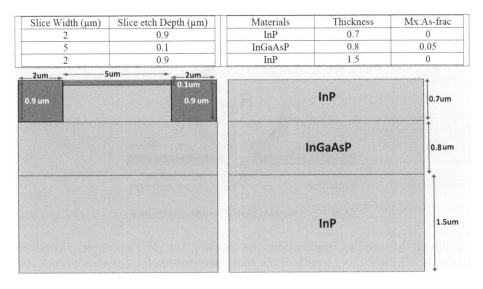

Figure 4.1 The waveguide section.

An often-used component in micro-ring-based optical circuitry is the directional coupler, which is a twin waveguide structure used to couple a fraction of light from one waveguide to another. Directional couplers were used to transfer light into and out of a MRR, and these can be designed with a high degree of accuracy.

4.2 PHYSICAL OF MICRO-RING RESONATORS

Optical ring waveguide resonators are useful components for wavelength filtering, multiplexing, switching and modulation. The main performance characteristics of these resonators are the Free-Spectral Range (FSR), the finesse (or -factor), the transmission at resonance, and the extinction ratio. The major physical characteristics underlying these performance criteria are the size of the ring, the propagation loss, and the input and output coupling ratios (such as reflectivity of a Fabry–Perot resonator). There are various components of losses, including sidewall scattering loss, bending radiation loss, and substrate leakage loss.

Critical to understanding how an optical ring resonator works, is the concept of how the linear waveguides are coupled to the ring waveguide. When a beam of light passes through a wave guide as shown in the graph on the right, part of light will be coupled into the optical ring resonator. The reason for this phenomenon to happen is because the wave property of the light, or if we consider it in ray optics, it is because of the transmission effect. In other words, if the ring and the waveguide are close enough, the light in the waveguide will be transmitted into the ring. There are three aspects that affect the optical coupling: the distance, the coupling length and the refractive indices between the waveguide and the optical ring resonator. In order to optimize the coupling, it is usually the case to narrow the distance between the ring resonator and the waveguide. The closer the distance, the easier the optical coupling happens.

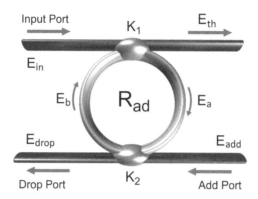

Figure 4.2 Schematic plot of an add-drop system.

In addition, the coupling length affects the coupling as well. The coupling length represents the effective curve length of the ring resonator for the coupling phenomenon to happen with the waveguide. It has been studied that as the optical coupling length increases, the difficulty for the coupling to happen decreases.

Furthermore, the refractive index of the waveguide material, the ring resonator material and the medium material in between the waveguide and the ring resonator also affect the optical coupling. The medium material is usually the important one been studied since it has a great effect on the transmission of the light wave. The refractive index of the medium can be either large or small according to various applications and purposes. Another critical issue in the structuring of micro resonators relates to the coupling coefficient between the waveguide and the resonator. To maximize the light extinction at the resonant wave-length, the coupling should be equal to the total propagation loss per resonator round trip. The horizontal coupling geometry requires a very accurate separation between the ring and the waveguide. To achieve a sub-micrometer gap, we lithographically define the waveguides and the rings in two steps, using a negative-tone photoresist (Poberaj *et al.*, 2009). In the first step the straight waveguides are created in the photoresist using mask photolithography and hardening. Subsequently, the rings are formed on a second photoresist layer with the same procedure and positioned using a standard mask-aligner. The two step technique, similar to the one presented in a recent work, reduces the diffraction effects that would inhibit the formation of the narrow gap if a single-step illumination was used.

A system of the ring resonator add-drop system is shown in Figure 4.2. The Gaussian pulse is introduced into the system as shown in this figure.

The input optical field (E_{in}) in the form of Gaussian pulse and add optical field (E_{add}) of the bright soliton pulse can be expressed by Equations (4.1) and (4.2).

$$E_{in}(z,t) = E_0 \exp\left[\left(\frac{z}{2L_D}\right) - i\omega_0 t\right] \tag{4.1}$$

$$E_{add}(z,t) = A \operatorname{sech}\left[\frac{T}{T_0}\right] \exp\left[\left(\frac{z}{2L_D}\right) - i\omega_0 t\right] \tag{4.2}$$

Here A and z are the optical field amplitude and propagation distance, respectively. T represents soliton pulse propagation time in a frame moving at the group velocity, $(T = t - \beta_1 \times z)$, where β_1 and β_2 are the coefficients of the linear and second order terms of the Taylor expansion of the propagation constant. The dispersion length of the soliton pulse can be defined as $L_D = T_0^2/|\beta_2|$, where the frequency carrier of the soliton is ω_0. Here, the soliton represents a pulse that keeps its width invariance as it propagates, known as a temporal or spatial soliton. The intensity of soliton peak is $(|\beta_2/\Gamma T_0^2|)$, where T_o is representing the initial soliton pulse propagation time. When a temporal soliton pulse propagates inside the micro-ring device, a balance should be achieved between the dispersion length (L_D) and the nonlinear length $(L_{NL} = 1/\Gamma \phi_{NL})$, where $\Gamma = n_2 \times k_0$, is the length scale over which disperse or nonlinear effects causes the beam becomes wider or narrower. For a soliton pulse when the balance between dispersion and nonlinear lengths is achieved, hence $L_D = L_{NL}$. The total refractive index (n) of the system is given by

$$n = n_0 + n_2 I = n_0 + \left(\frac{n_2}{A_{\text{eff}}}\right) P, \tag{4.3}$$

where n_0 and n_2 are the linear and nonlinear refractive indices, respectively. I and P are the optical intensity and optical power, respectively. A_{eff} represents the effective mode core area of the device, where in the case of MRRs, the effective mode core areas range from 0.50 to 0.25 μm^2. When a Gaussian pulse is input and propagates within the MRR, the resonant output is formed for each round-trip. The normalized output of the light field is defined as the ratio between the output and input fields ($E_{\text{out}}(t)$ and $E_{\text{in}}(t)$) in each round-trip. Thus, it can be expressed as Equation (4.4).

$$\frac{E_{\text{out}}(t)}{E_{\text{in}}(t)} = \sqrt{(1 - \gamma) \times \left[1 - \frac{(1 - (1 - \gamma) x^2) \kappa_1}{(1 - x\sqrt{1 - \gamma}\sqrt{1 - \kappa_1})^2 + 4x\sqrt{1 - \gamma}\sqrt{1 - \kappa_1}\sin^2\left(\frac{\phi}{2}\right)}\right]} \tag{4.4}$$

Here, κ is the coupling coefficient, $x = \exp(-\alpha L/2)$ represents a round-trip loss coefficient, $\phi_0 = kLn_0$ and $\phi_{NL} = kLn_2|E_{\text{in}}|^2$ are the linear and nonlinear phase shifts and $k = 2\pi/\lambda$ is the wave propagation number. Here L and α are the waveguide length and linear absorption coefficient, respectively. To retrieve the signals from the chaotic noise, we propose to use an add-drop interferometer system with appropriate parameters. The input powers expressed by Equations (4.1) and (4.2), insert into the input and add ports of the add-drop system. Interior optical signals of the system can be expressed by Equations (4.5) and (4.6).

$$E_a = \frac{E_{\text{in}} \times j\sqrt{\kappa_1} + E_{\text{add}} \times j\sqrt{\kappa_2 \times (1 - \kappa_1)} \times e^{-\alpha \frac{L_{\text{ad}}}{2} \cdot \frac{1}{2} - jk_n \frac{L_{\text{ad}}}{2}}}{1 - \sqrt{(1 - \kappa_1) \times (1 - \kappa_2)} \times e^{-\frac{\alpha L_{\text{ad}}}{2} - jk_n L_{\text{ad}}}} \tag{4.5}$$

$$E_b = \frac{E_{\text{in}} \times j\sqrt{\kappa_1 \times (1 - \kappa_2)} \times e^{\frac{-\alpha}{2}\frac{L_{\text{ad}}}{2} - jk_n\frac{L_{\text{ad}}}{2}}}{1 - \sqrt{(1 - \kappa_1) \times (1 - \kappa_2)} \times e^{\frac{-\alpha L_{\text{ad}}}{2} - jk_n L_{\text{ad}}}}$$

$$+ \frac{E_{\text{add}} \times j\sqrt{\kappa_2 \times (1 - \kappa_1) \times (1 - \kappa_2)} \times e^{\frac{-\alpha L_{\text{ad}}}{2} - jk_n L_{\text{ad}}}}{1 - \sqrt{(1 - \kappa_1) \times (1 - \kappa_2)} \times e^{\frac{-\alpha L_{\text{ad}}}{2} - jk_n L_{\text{ad}}}} + E_{\text{add}} \times j\sqrt{\kappa_2}, \tag{4.6}$$

where κ_1 and κ_2 are the coupling coefficients, $L_{\text{ad}} = 2\pi R_{\text{ad}}$ and R_{ad} is the radius of the add-drop interferometer system. The through and drop ports output signals from the system are given by:

$$E_{\text{th}} = E_b \times j\sqrt{\kappa_1} \times e^{\frac{-\alpha}{2}\frac{L_{\text{ad}}}{2} - jk_n\frac{L_{\text{ad}}}{2}} + E_{\text{in}} \times \sqrt{1 - \kappa_1} \tag{4.7}$$

$$E_{\text{drop}} = E_a \times j\sqrt{\kappa_2} \times e^{\frac{-\alpha}{2}\frac{L_{\text{ad}}}{2} - jk_n\frac{L_{\text{ad}}}{2}} + E_{\text{add}} \times \sqrt{1 - \kappa_2}, \tag{4.8}$$

where E_{th} and E_{drop} represent the optical electric fields of the through and drop ports, respectively. Therefore,

$$E_{\text{th}} = \frac{-E_{\text{in}} \times \kappa_1 \times \sqrt{1 - \kappa_2} \times e^{\frac{-\alpha L_{\text{ad}}}{2} - jk_n L_{\text{ad}}}}{1 - \sqrt{(1 - \kappa_1) \times (1 - \kappa_2)} \times e^{\frac{-\alpha L_{\text{ad}}}{2} - jk_n L_{\text{ad}}}}$$

$$+ \frac{-E_{\text{add}} \times \sqrt{(\kappa_1 \times \kappa_2) \times (1 - \kappa_1) \times (1 - \kappa_2)} \times e^{\frac{-3\alpha L_{\text{ad}}}{4} - jk_n\frac{3L_{\text{ad}}}{2}}}{1 - \sqrt{(1 - \kappa_1) \times (1 - \kappa_2)} \times e^{\frac{-\alpha L_{\text{ad}}}{2} - jk_n L_{\text{ad}}}} \tag{4.9}$$

$$- E_{\text{add}} \times \sqrt{\kappa_1 \times \kappa_2} \times e^{\frac{-\alpha L_{\text{ad}}}{4} - jk_n\frac{L_{\text{ad}}}{2}} + E_{\text{in}} \times \sqrt{1 - \kappa_1}$$

$$E_{\text{drop}} = \frac{-E_{\text{in}} \times \sqrt{\kappa_1 \times \kappa_2} \times e^{\frac{-\alpha}{2}\frac{L_{\text{ad}}}{2} - jk_n\frac{L_{\text{ad}}}{2}}}{1 - \sqrt{(1 - \kappa_1) \times (1 - \kappa_2)} \times e^{\frac{-\alpha L_{\text{ad}}}{2} - jk_n L_{\text{ad}}}}$$

$$- \frac{E_{\text{add}} \times \kappa_2 \times \sqrt{1 - \kappa_1} \times e^{\frac{-\alpha}{2}L_{\text{ad}} - jk_n L_{\text{ad}}}}{1 - \sqrt{(1 - \kappa_1) \times (1 - \kappa_2)} \times e^{\frac{-\alpha L_{\text{ad}}}{2} - jk_n L_{\text{ad}}}} + E_{\text{add}} \times \sqrt{1 - \kappa_2} \tag{4.10}$$

The waveguide (ring resonator) loss is $\alpha = 0.5$ dBmm^{-1}, where the fractional coupler intensity loss is $\gamma = 0.1$. The microscopic image of an add-drop system with radius of $635\,\mu$m is shown in Figure 4.3.

The filter response of the MRR add-drop system with two waveguides and coupling factor of $\kappa_1 = \kappa_2 = 0.15$, $R = 250\,\mu$m and $\alpha = 0$ in both symmetrical couplers is shown in Figure 4.4.

In the following new parameter will be used for simplification:

$$D = (1 - \gamma)^{\frac{1}{2}} x = D \exp\left(-\frac{\alpha}{2}L\right), \quad y_1 = \sqrt{1 - \kappa_1} \quad \text{and} \quad y_2 = \sqrt{1 - \kappa_2}$$

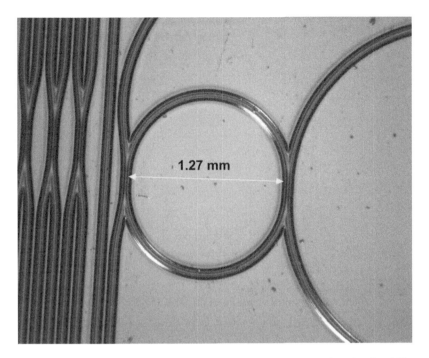

Figure 4.3 Microscopic image of waveguide MRR add-drop system with radius $R = 635\,\mu$m.

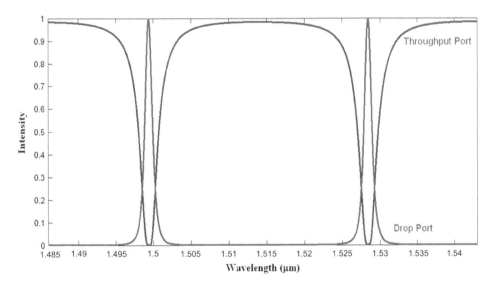

Figure 4.4 Transmission feature of the MRR add-drop system.

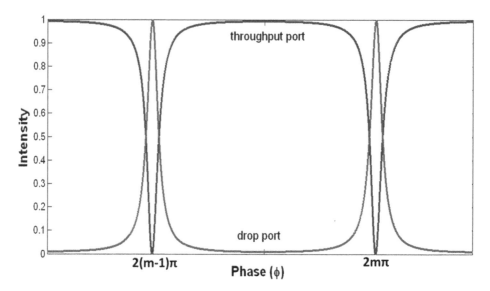

Figure 4.5 Transmission feature of the MRR add-drop ring system.

The maximum and minimum transmission are calculated as follows. For the throughput port:

$$T_{max} = \frac{(y_1 + y_2 x)^2}{(1 + y_1 y_2 x)^2} \tag{4.11}$$

$$T_{min} = \frac{(y_1 - y_2 x)^2}{(1 - y_1 y_2 x)^2}, \tag{4.12}$$

and for the drop port:

$$T_{max} = \frac{(1 - y_1^2)(1 - y_2^2)x}{(1 - y_1 y_2 x)^2} \tag{4.13}$$

$$T_{min} = \frac{(1 - y_1^2)(1 - y_2^2)x}{(1 - y_1 y_2 x)^2} \tag{4.14}$$

The output intensity at the throughput port will be zero at resonance ($k_n L = 2m\pi$) shown in Figure 4.5, which indicates that the resonance wavelength is fully extracted by the resonator.

The insertion loss in dB can be given as Equation (4.15), where the powers transmitted and received are presented by P_T and P_R respectively.

$$IL = 10 \log_{10} \frac{P_T}{P_R} \tag{4.15}$$

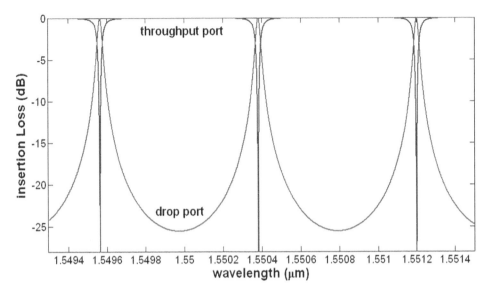

Figure 4.6 Insertion loss of the MRR add-drop system.

The insertion loss profile of the MRR add-drop system is shown in Figure 4.6.

MRRs are expected to be essential components in next-generation, integrated photonic circuits. The light coupled into the resonator via a waveguide is confined within the MRR cavity due to total internal reflections and high-Q resonant modes are formed. The positions of these modes depend on the effective index of the resonant structure and thus get shifted when there is a molecular interaction on the surface. This shift can be determined with high precision using our method of detection. To selectively excite or suppress longitudinal modes of MRRs for broader-band operation, major effort has gone into engineering optical ring cavities, e.g., using series-coupled or cascaded multiple rings with the Vernier effect or inserting Bragg gratings inside the ring cavity. Control of the couplings between the MRRs and the bus waveguides is also critical for shaping the spectral responses, which is challenging when the bend radius is scaled down to a few micrometers. Optical MRRs based filters and modulators are the potential building blocks of very-large-scale integration (VLSI) photonic circuits. The ratio of optical power coupled between ring and bus waveguides in MRR based devices is crucial to the desired device characteristics, and is the subject of study in the integrated MRR photonics devices.

Chapter 5

Micro-ring modulators

5.1 INTRODUCTION

Growing bandwidth needs have presented the need for optical communication at scales and distances smaller than previously precedented, motivating the use of optical links for such scenarios as rack-to-rack links in data centers, board-to-board interconnects, and ultimately for use in multi-core processors (Savage, 2002; Pepeljugoski *et al.*, 2010). However, at these smaller scales, optical links are only feasible if they can be realized in a small footprint and energy-efficient manner. For this reason, the silicon photonics platform, with its ability to manifest CMOS-compatible photonic devices, is promising for use in next-generation optical links (Liu *et al.*, 2010; Reed *et al.*, 2010). Within the silicon photonics platform, metrics of size and energy efficiency (Padmaraju & Bergman, 2013; Padmaraju *et al.*, 2013).

However, as the high-performance functionality of both passive and active silicon micro-ring-based devices have continued to be demonstrated, concerns have grown over the suitability of these devices for use in thermally volatile environments (Ahn *et al.*, 2009). The high thermo-optic coefficient of silicon, combined with the resonant nature of the micro-ring-based devices, lends the operation of said devices susceptible to thermal fluctuations of only a few kelvin (K) (Kawachi, 1990; Cocorullo *et al.*, 1999). Attempts to resolve the thermal sensitivity of passive and active micro-ring-based devices have been focused on creating thermally insensitive structures or dynamic feedback systems (Padmaraju *et al.*, 2012a; Padmaraju *et al.*, 2012b). In particular, it has been shown that using a feedback system, a micro-ring modulator can maintain error-free performance under thermal fluctuations that would normally render it inoperable (Padmaraju *et al.*, 2012b; Padmaraju *et al.*, 2013). A feedback system thermally stabilizes the micro-ring modulator by monitoring the temperature, either directly or indirectly, and then adjusting the local temperature of the modulator using an appropriate mechanism. In the cited demonstration, changes in the temperature of the micro-ring modulator were inferred by monitoring the mean power of the micro-ring-modulated signal using an off-chip photodetector (Nixon *et al.*, 1995; Guidash *et al.*, 1997). To complete the feedback system, the bias current was varied to provide the necessary temperature adjustment in the localized region of the micro-ring modulator (Razavi, 1996; Perić *et al.*, 2006).

Optical interconnections on in recent years as interconnections become the bottleneck for the next-generation computing systems (Lippmaa *et al.*, 1981; Peng *et al.*, 2002). The goal of these investigations is to provide a compact, low-power-consumption, high bandwidth and low-latency optical interconnection system with full CMOS-compatibility. Silicon based optical components such as low-loss Silicon-On-Insulator (SOI) optical waveguides, high-speed silicon modulators, and Ge-on-SOI detectors have been demonstrated, enabling large-scale optical integration on a silicon chip (Liu *et al.*, 2004, 2007). While most attention is focused on single-channel systems at this stage, Wavelength Division Multiplexing (WDM) technology is necessary to fully utilize the ultra-wide bandwidth of the optical interconnection medium, given that the transmission bandwidths of both silicon waveguides and optical fibers are on the order of 10–100 THz (Keiser, 2003; Baehr-Jones *et al.*, 2005). In this book, we present a simple architecture for a WDM interconnection system based on silicon ring resonators. As the key components of such a system, we show high-speed and multi-channel modulation using cascaded silicon micro-ring modulators (Xu *et al.*, 2008).

5.2   MICRO-RING USED AS MODULATOR

The silicon modulator is a key component for CMOS-compatible optical interconnection systems (Dong *et al.*, 2009). Recently, high-speed modulators based on free-carrier plasma dispersion effect have been demonstrated using either (MZI) or Micro-Ring Resonators (MRRs) in (Mårtensson *et al.*, 2004; Liu *et al.*, 2007; Li & Tong, 2008). Comparing to the mm-long MZI-based modulators, the advantages of the ring-resonator-based modulator include its small size (~10 µm) and low-power consumption (McCall *et al.*, 1992; Roundy *et al.*, 2003). In the ring resonator, in contrast to single-pass devices like MZI, light at the resonant wavelength travels many round trips in the resonator, and interacts with the carriers many times (Chao *et al.*, 2007). As a result, the total number of carriers needed to change the optical transmission of the ring resonator is much less than that needed in MZI-modulators, and therefore much less RF power is needed to drive these carriers in and out of the active region (Henry, 1982).

For the WDM interconnection systems, in addition to the low power consumption and small size, ring modulators have another advantage: they modulate only light at particular wavelengths (the resonant wavelengths of the ring resonators) and allow light at all other wavelengths to pass through the modulators without been affected (Keeler *et al.*, 2003). Therefore, one can cascade several ring modulators with different resonant wavelengths on a single waveguide, and modulate different wavelengths of light independently. Fig. 1 shows a simple architecture for a WDM interconnection system (Xu *et al.*, 2005; Xia *et al.*, 2006). A similar structure has been proposed for polymer modulators (Chen *et al.*, 1997). Light from a WDM source (Kim *et al.*, 2000) or a broadband source (Provino *et al.*, 2001) is sent into a silicon waveguide coupled to multiple ring modulators with different resonant wavelengths. If the input is a WDM source, the resonant wavelength of each modulator needs to match the wavelength of each channel of the WDM source. At the receiver side, these channels can be with drop ports, and detected separately. If the input is a broadband source, it only requires that the resonant wavelengths of the ring modulators match one-to-one with those of the ring de-multiplexers (Oda *et al.*, 1988).

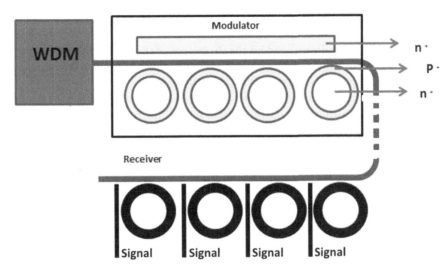

Figure 5.1   Schematics of a WDM optical interconnection system with cascaded silicon ring resonators as a WDM modulator and demultiplexer.

The key components of the WDM interconnection system are the cascaded modulators shown in the shadowed area of Figure 5.1, which are fabricated on a SOI substrate. The device structure is based on the micro-ring modulator. They consist of ring resonators embedded with PIN junctions used to inject and extract free carriers, which in turn modify the refractive index of the silicon and the resonant wavelength of the ring resonator using the mechanism of the plasma dispersion effect (Djordjev *et al.*, 2002a). The waveguides and rings are formed by silicon strips. The speed of the modulator was limited to 400 Mbps under Non-Return-to-Zero (NRZ) coding (Possley & Upham, 2010). The reason for this limitation is that the p-i-n junction is formed on only part of the ring resonator, while carriers diffuse into the section of the ring that is not part of the p-i-n junction, where they cannot be efficiently extracted during the reverse biased period, leading to a longer fall time following consecutive '1's. In the new design presented here, an additional n^+-doped region is added outside of the straight waveguide to form nearly closed p-i-n junctions. This new geometry ensures that all carriers injected into the ring can be extracted efficiently by reversely biasing the junction. The distance between the doped regions and the edge of the ring resonators and straight waveguides is reduced to nanometer range to further increase the extraction speed with the same reverse bias voltage. The radii of the four ring resonators are designed to be sub-micrometers. The difference in the radii corresponds to a channel spacing. A top-view microscopic picture of two of the four fabricated ring modulators is shown in Figure 5.1. The image shows both the ring resonators coupled to the straight waveguide and the metal pads contacting the doped regions.

In this demonstration, we further the results of by showing that the optoelectronic components that comprise the feedback system can be integrated onto a single device using CMOS-compatible processes and materials (Burdea & Langrana, 1995).

The enabling technology of this integration is the use of a defect-enhanced silicon photodiode (Kempf, 2005). Such devices have been demonstrated as effective high-speed optical receivers, but an additional utility lies in their use as in-situ power monitors for silicon photonic devices. Positioned on the drop-port of the micro-ring modulator, the silicon photodiode is utilized as the photoreceiver needed to monitor the mean power of the modulated signal (Padmaraju *et al.*, 2013).

This configuration avoids the use of a power tap and is compatible with the WDM arrangement of micro-ring modulators, where several micro-rings are cascaded along the same waveguide bus.

5.3 FREQUENCY-DEPENDENT MICRO-RING TRANSMISSION

We will start by describing the frequency-dependent transmission of the ring modulator, the basis for the time-domain quasi static model of INTERCONNECT. The resonator modulator configuration is depicted in Figure 5.2. A Continuous Wave (CW) incoming optical wave is modulated by varying the refractive index of the MRR. Different from Fabry-Perot resonators, ring resonators use couplers instead of mirrors for optical feedback.

Where A is the input optical field at port 1, and B is the output optical field (through) at port 2. a is the attenuation after each round-trip, t_1, k_1 and t_2, k_2 are the resonator transmission and coupling coefficients for the first and second coupler respectively.

5.4 OPTICAL MODULATOR BASED ON MRRS INTEGRATED WITH MACH-ZEHNDER INTERFEROMETER (MZI)

Recent innovations and breakthroughs in silicon photonics are paving the way for the realization of high speed on-chip optical interconnects. Transfer of information between components requires that data be superimposed on the optical carrier signal

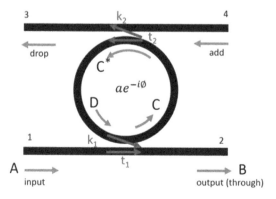

Figure 5.2 Schematic representation of a ring resonator modulator device.

by electro-optic modulation. Numerous high performance silicon electro-optic modulators have been demonstrated which generate Non-Return-to-Zero (NRZ) encoding at bit-rates as high as 40 Gbps. However, there are numerous other optical modulation formats which could yield improved performance of the optical links such as better Signal/Noise ratio, reduced nonlinearity or even higher bit-rates.

Some recent examples of alternate encodings on a silicon photonic platform are the use of MRRs to convert Non-Return-to-Zero (NRZ) to Pseudo-Return-to-Zero (PRZ) in order to aid clock recovery and the generation of Return-to-Zero-Differential-Phase-Shift-Keying (RZ-DPSK) signals with improved chirp. Here, we propose a scheme for generating Amplitude-Shift-Keying (ASK) format in order to significantly increase the bit-rate of on-chip optical links by using FDTD method. As for the numerical simulation of all-optical ASK-to-PSK, all our numerical work has been carried out by using commercially available simulation software-the Opti-FDTD simulation package.

Here, three amplitude level signals can be generated using a triple of symmetric MRRs arranged in series in a Mach-Zehnder Interferometer (MZI) configuration as seen in Figure 5.3. The device works by splitting the input light into two separate paths with a 3-dB coupler. When the light is on resonance with the MRRs, it is coupled to the out1 ports where it constructively interferes at the output port. If one MRR is shifted off-resonance, the output of the system is halved because only half of the light transfers to the out2 port as illustrated in Figure 5.3.

This system can be used to amplitude generate using single nonlinear SMRR, double nonlinear MRRs (DMRRs), and triple nonlinear MRRs (TMRRs) coupled to one arm of MZI. We have found that the enhanced amplitude is increased when TMRRs are placed into the system. The ASK generation can be obtained using SMRR, where the upper output (out1) and lower output (out2) Amplitude Shift Keying (ASK) occurred, and Phase Shift Keying (PSK) enhancement is equal to π. When the number of nonlinear MRRs increases, for instance, MZI arm is coupled by DMRRs and TMRRs, the ASK enhanced peak is increased, where the PSK enhanced peak remain equal to π.

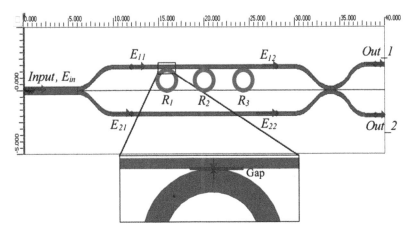

Figure 5.3 Scheme of ASK-to-PSK in InGaAsP/InP waveguide size 40 μm².

Transfer of information between components requires that data be superimposed on the optical carrier signal by electro-optic modulation. Therefore, the ASK modulation can be applied based on the SMRR, DMRRs and TMRRs, which is coupled into one arm MZI. It can be found that the number of ASK enhanced peaks will increase more than number of nonlinear MRRs. In order to generate three amplitude levels, all nonlinear MRRs should be modulated simultaneously. When the TMRRs is in resonance, a signal as "1" can be generated, and when one of the MRRs is off-resonance, then a signal as "0" can be generated. Therefore, it is performed that with TMRRs, it is possible to generate up to three different amplitude levels by coupling integrated MRRs into the MZI, results on build up an optical device. This system can merely be scaled up to more logic levels by adding extra nonlinear MRRs or optical splitters to the system.

Chapter 6

Micro-Ring Resonator (MRR) in optical transmission systems

6.1 MICRO-RING RESONATOR SYSTEMS IN OPTICAL COMMUNICATION SYSTEMS

The solitons have been extensively investigated in many physics studies. The fundamental mechanism of soliton formation, namely, the balanced interplay of linear GVD and nonlinearity-induced Self-Phase Modulation (SPM), is a well-understood concept. The soliton pulses are so stable that the shape and velocity are preserved while travelling along the medium. An optical soliton pulse is recommended in order to create a spectrum of light over a wide range, where they are powerful laser pulses that can be employed to generate chaotic filter characteristics. Generation of multi solitons becomes an interesting subject when it is used to enlarge the capacity of communication channels. The system of nonlinear ring resonators can be used to generate chaotic signals. In this chapter, we propose a modified add/drop optical filter called PANDA system that consists of one centered ring resonator connected to two smaller ring resonators on the right and left sides. By controlling some suitable parameters, the generated result within the ring resonator system can be controlled.

The chaotic signals generated by the PANDA system can be transmitted within an optical fiber transmission link, where the multi transmitted ultra-short spatial and temporal solitons can be generated. The Generation of multi soliton pulses has become an interesting approach to enlarging communication channel capacity. The dynamics of ultra-short pulse propagation in a MRR system have recently attracted research interest because such pulses are characterized by wide bandwidths and high speeds. One main application of the multiple soliton transmission system, which is made of integrated ring resonators, is a high data-rate transmission for short and long distance communications. Besides improvements in efficiency and beam quality, these generated multi soliton provide short and ultra-short bandwidth, leading to improved process efficiencies and new fields of laser applications. The main advantage of using soliton pulses in optical communication systems is that the shape of the pulse remains almost unaltered over a long distance.

In order to improve the system, narrower soliton pulses are recommended, where the attenuation of such signals during transmission lessens when compared to the conventional peaks of micrometre laser pulses. Generated optical pulses can be used for optical communications, in which the capacity of the output signals can be improved through the generation of peaks with smaller FWHMs. To transmit the soliton signals via long distance communications, ultra-short soliton pulses are required. In this study

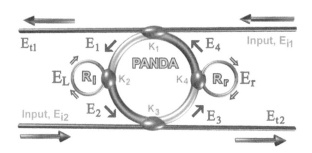

Figure 6.1 Schematic diagram of a PANDA ring resonator system.

chaotic signals in the form of logic codes are generated by the PANDA system and are transmitted via a fiber optic transmission link with the length of 180 km, where the non-linear behavior of the fiber causes the signals to be compressed along the transmission link. This technique provides spatial and temporal soliton pulses with an ultra-short bandwidth of picometers and picoseconds.

In this chapter, a passive MRR system known as a PANDA ring resonator is presented. This system is used to generate signals of the type of solitons. The intense chaotic signals can be generated and transmitted within a nonlinear Kerr medium by using the resonant conditions. Using additional Gaussian pulses input to the add port of the ring system, the results of chaotic signals can be controlled in a proper way. A balance should be achieved between dispersion and nonlinear lengths in the case of soliton propagations. The chaotic output signals generated by the ring resonator are converted to codes then inserted into an optical transmission link in order to perform the transmission process. The receiver is used to detect the transmitted signals and show received signals in the form of spatial and temporal solitons.

6.2 THEORETICAL BACKGROUND OF SOLITON PROPAGATION IN NONLINEAR KERR MEDIUM

The proposed system of chaotic signal generation is known as a PANDA ring resonator (Figure 6.1), where two input signals of Gaussian laser beam can be introduced into the system via the input and add ports.

The Kerr effect causes the refractive index (n) of the medium to be varied and it is given by

$$n = n_0 + n_2 I = n_0 + \frac{n_2}{A_{\text{eff}}} P \tag{6.1}$$

With n_0 and n_2 as the linear and nonlinear refractive indexes, respectively. I and P are the optical intensity and the power, respectively. The effective mode core area of the device (A_{eff}) ranges from 0.50 to 0.10 μm^2. Input optical fields of Gaussian pulses are given by

$$E_{i1}(t) = E_{i2}(t) = E_0 \exp\left[\left(\frac{z}{2L_D}\right) - i\omega_0 t\right], \tag{6.2}$$

E_0 and z are the amplitude of optical field and propagation distance respectively. L_D is the dispersion length where frequency shift of the signal is ω_0. The electric field of the left ring of the PANDA system is given by:

$$E_L = \left(E_1\sqrt{1-\gamma_2}\right) \times \frac{\sqrt{1-\kappa_2} - \sqrt{1-\gamma_2}e^{-\frac{\alpha}{2}L_L - jk_n L_L}}{1 - \sqrt{(1-\gamma_2)(1-\kappa_2)}e^{-\frac{\alpha}{2}L_L - jk_n L_L}}. \tag{6.3}$$

κ is the intensity coupling coefficient, $k = 2\pi/\lambda$ is the wave propagation, γ is the fractional coupler intensity loss, $L_L = 2\pi R_l$, R_l is the radius of left ring. The electric field of the right ring of the PANDA system is given as:

$$E_r = \left(E_3\sqrt{1-\gamma_4}\right) \times \frac{\sqrt{1-\kappa_4} - \sqrt{1-\gamma_4}e^{-\frac{\alpha}{2}L_R - jk_n L_R}}{1 - \sqrt{1-\gamma_4}\sqrt{1-\kappa_4}e^{-\frac{\alpha}{2}L_R - jk_n L_R}}, \tag{6.4}$$

Here, $L_R = 2\pi R_r$ and R_r is the radius of right ring. We define the parameters of x_1, x_2, y_1 and y_2 as: $x_1 = (1-\gamma_1)^{\frac{1}{2}}$, $x_2 = (1-\gamma_3)^{\frac{1}{2}}$, $y_1 = (1-\kappa_1)^{\frac{1}{2}}$, and $y_2 = (1-\kappa_3)^{\frac{1}{2}}$, thus the interior signals can be expressed by,

$$E_1 = \frac{jx_1\left[\sqrt{\kappa_1}E_{i1} + x_2 y_1 \sqrt{\kappa_3}E_r E_{i2}e^{-\frac{\alpha L}{4} - jk_n \frac{L}{2}}\right]}{1 - x_1 x_2 y_1 y_2 E_L E_r e^{-\frac{\alpha}{2}L - jk_n L}}, \tag{6.5}$$

$$E_2 = E_L E_1 e^{-\frac{\alpha L}{4} - jk_n \frac{L}{2}}, \tag{6.6}$$

$$E_3 = x_2\left[y_2 E_L E_1 e^{-\frac{\alpha L}{4} - jk_n \frac{L}{2}} + j\sqrt{\kappa_3}E_{i2}\right], \tag{6.7}$$

$$E_4 = E_r E_3 e^{-\frac{\alpha L}{4} - jk_n \frac{L}{2}}. \tag{6.8}$$

L is the circumference of the PANDA ring resonator. Output electric fields of the PANDA system given by E_{t1} and E_{t2} and are expressed as:

$$E_{t1} = AE_{i1} - \frac{G^2 B E_{i2}e^{-\frac{\alpha L}{4} - jk_n \frac{L}{2}}}{1 - FG^2}[CE_{i1} + DE_{i2}G], \tag{6.9}$$

$$E_{t2} = \frac{Gx_2 y_2 E_{i2}\sqrt{\kappa_1 \kappa_3}}{1 - FG^2}\left[AE_L E_{i1} + \frac{D}{x_1 \kappa_1 \sqrt{\kappa_3}E_r}E_{i2}G\right], \tag{6.10}$$

where, $A = x_1 x_2$, $B = x_1 x_2 y_2 \sqrt{\kappa_1}E_r$, $C = x_1^2 x_2 \kappa_1 \sqrt{\kappa_3}E_L E_r$, $G = \left(e^{-\frac{\alpha L}{4} - jk_n \frac{L}{2}}\right)$, $D = (x_1 x_2)^2 y_1 y_2 \sqrt{\kappa_1 \kappa_3}E_L E_r^2$ and $F = x_1 x_2 y_1 y_2 E_L E_r$.

6.3 RESULT AND DISCUSSION

Gaussian beams with a central wavelength of $1.55\,\mu$m and a power of $0.6\,$W are introduced into the add and input ports of the PANDA ring resonator. The fiber system exhibits a nonlinear Kerr effect, where the linear and nonlinear refractive indices of the

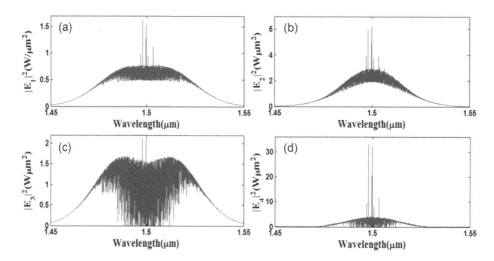

Figure 6.2 Interior signal generation in the PANDA ring resonator system, where (a): $|E_1|^2$, (b): $|E_2|^2$, (c): $|E_3|^2$ and (d): $|E_4|^2$.

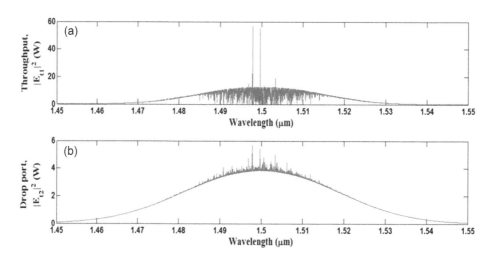

Figure 6.3 Chaotic signal generation using the PANDA system where (a): Throughput chaotic signals and (b): Drop port chaotic signals.

system are $n_0 = 3.34[64, 65]$ and $n_2 = 1.3 \times 10^{-17}$, respectively. The selected radius of the cantered ring resonator is $R_{\text{PANDA}} = 300$ nm, where the right and left ring resonators have radii of 180 and 200 nm, respectively. The simulated interior signals from the PANDA system are shown in Figure 6.2. The coupling coefficients of the PANDA ring resonator are $\kappa_1 = 0.35$, $\kappa_2 = 0.2$, $\kappa_3 = 0.1$, and $\kappa_4 = 0.95$.

More channel capacity can be obtained and controlled by generating large bandwidth of chaotic signals. Therefore, stable signals of the chaotic signals can

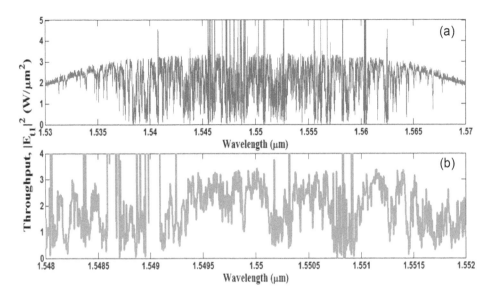

Figure 6.4 Chaotic signal generation using the PANDA system where (a): Throughput chaotic signals and (b): Expansion of the throughput chaotic signals.

Figure 6.5 Randomly generated logic codes within the chaotic signals with minimum and maximum intensity power of 2 and 2.4 W/μm^2.

be seen within the through and drop ports of the system shown in Figure 6.3. Figure 6.3 shows the through port chaotic signals where the expansion of the signals can be seen in Figure 6.3(b). The generated chaotic signals are distributed over the wavelength ranges from 1.53 μm to 1.57 μm. These types of signals can be used as carrier signals, where information can be carried out by the signals via an optical communication link. In order to transmit the signals, a fiber optic transmission link can be used, therefore multi soliton pulses can be generated and used in many applications in optical communications. Generated logic code is as "00101000100100000101101101100010010111011011101001" within the chaotic signals can be shown by Figure 6.5.

The potential of multi soliton pulses can be used for many applications such as high capacity and secured optical communication. Thus, the chaotic signals from the

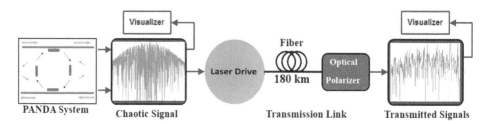

Figure 6.6   Optical transmission link, where the fiber optic has a length of 180 km.

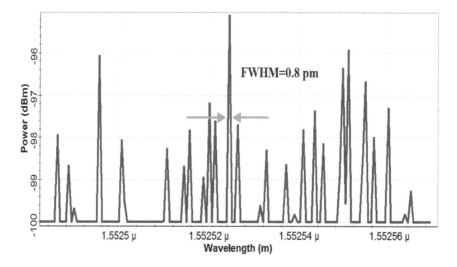

Figure 6.7   Spatial multi solitons with FWHM = 0.8 pm.

through port of the system in the form of codes can be input into the fiber optic transmission link to perform the optical quantum transmission process. The transmission link system is shown in Figure 6.6.

In Figure 6.6, the fiber optic has a length of 180 km, attenuation of 0.4 dB/km, dispersion of 1.67 ps/(nm.km), the differential group delay of 0.2 ps/km, the nonlinear refractive index of 2.6×10^{-20} m^2/W, effective area of 50 μm^2 and the nonlinear phase shift of 3 m rad. Figure 6.7 shows the transmitted chaotic signals in the communication system, which leads to generate spatial multi solitons. The generation of ultra-short dark and bright soliton signals can be obtained after the chaotic signals were transmitted along the fiber optic transmission link, where finally the signals are received by suitable optical receiver thus the detection process can be performed via the optical receiver. The FWHM of the spatial multi soliton signals is 0.8 pm. The temporal shape of the multi soliton pulses can be seen in Figure 6.8. Here the temporal pulses with FWHM of 60 ps could be generated experimentally.

Therefore, transmission of the chaotic signals along the fiber optic is performed, where the spatial and temporal solitons can be generated and detected using a suitable optical receiver.

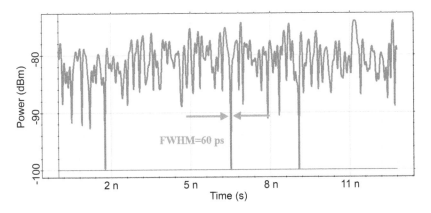

Figure 6.8 Temporal multi soliton pulses with FWHM of 60 ps.

In conclusion, the PANDA is presented as optical chaos. The Gaussian beams with central wavelengths of 1.55 μm and powers of 0.6 W are inserted into the PANDA system, where a high capacity of chaotic signals can be generated. In order to compress the noisy chaotic signals, we transmit these signals in the form of codes via an optical fiber optic transmission link with the length of 180 km. At the end of the transmission link, the clear and filtered signals of spatial and temporal solitons can be generated and used for many applications in optical communications. Here the spatial and temporal signals with FWHM of 0.8 pm and 60 ps could be generated respectively as shown in Figure 6.8.

Chapter 7

Methods of slow light generation

7.1 INTRODUCTION

This chapter presents the techniques of generating slow and fast pulse using micro-ring resonator. Micro-ring resonator systems made of different materials. Micro-ring device can be constructed with radius in the size of a micron. The concepts behind optical ring resonators are the same as those behind whispering galleries except that they use light and obey the properties behind constructive interference and total internal reflection. In this research, MRRs are made of InGaAsP/InP, GaAlAs/GaAs and hydrogenated amorphous silicon (a-Si:H) materials. The important effects of these materials are dispersion and nonlinearity in optical waveguide while the pulse fed into MRR. Each optical ring resonator behaves in many ways like a Fabry-Perot cavity.

Fabrication of InGaAs/InP waveguide is based on the semiconductor materials. In designing material and device structures for waveguide modulators, one has to consider various physical constraints limiting the microwave intensity, output power, the modulation depth, and the bandwidth. The capacitance of the modulator must be minimized for wide bandwidth consideration.

Recently, a variety of GaAs/AlGaAs waveguide devices have been reported, including directional couplers, MRRs and photonic crystal cavities.

Interest in hydrogenated amorphous silicon as a material for the realization of optical interconnects in integrated circuits was initially proposed with the demonstration of an optical tunable a-Si:H planar waveguide based Fabry-Perot intensity modulator. The a-Si:H is an alternative material which can be used for integration of silicon photonics. It allows amorphous silicon to be integrated at any point in the fabrication process with minimal complexity enabling vertical stacking of optical interconnects. Low loss waveguides including cavity resonators have been demonstrated using amorphous silicon (Harke et al., 2005).

The characteristics of waveguide materials are listed in Table 7.1 for center wavelength 1.5 μm (Paul et al., 1992). Figure 7.1 shows the flowchart of computing the output power, intensity, FWHM, delay time, delay phase and generally slow and fast light generation using linear and nonlinear waveguide.

In optical communication systems, the term nonlinearity refers to the dependence of the system on power of the optical beams being launched into the fiber cable. Nonlinear effects in optical fibers have become an area of academic research and of great importance in the optical fiber based systems. Several experiments in the past

Table 7.1 Properties of used waveguide material.

Properties	InGaAsP/InP	GaAlAs/GaAs	a-Si:H
Core refractive index	3.34	3.37	3.48
Clad refractive index	3.17	3.14	3.1
Core area	0.1–0.9 μm²	0.085–0.9 μm²	0.1–0.9 μm²
Nonlinear refractive index	$3.2 \times 10^{-17} m^2 W^{-1}$	$5.4 \times 10^{-18} m^2 W^{-1}$	$4.2 \times 10^{-17} m^2 W^{-1}$
Absorption	20 dB/cm	15 dB/cm	13.5 dB/cm
Center Wavelength	1.5 μm	1.5 μm	1.5 μm
Coupling loss	0.01	0.01	0.01

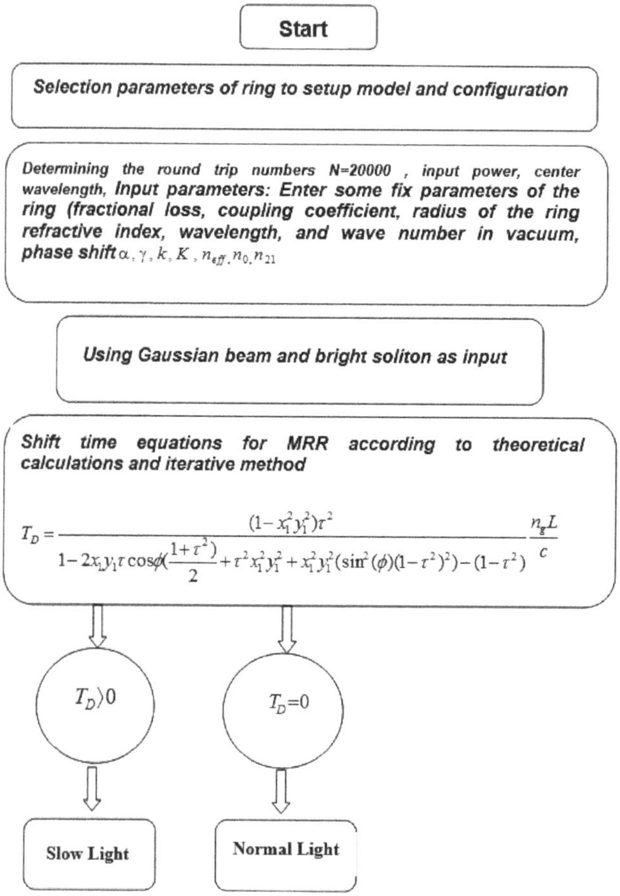

Start

Selection parameters of ring to setup model and configuration

Determining the round trip numbers N=20000 , input power, center wavelength, Input parameters: Enter some fix parameters of the ring (fractional loss, coupling coefficient, radius of the ring refractive index, wavelength, and wave number in vacuum, phase shift $\alpha, \gamma, k, K, n_{eff}, n_0, n_{21}$

Using Gaussian beam and bright soliton as input

Shift time equations for MRR according to theoretical calculations and iterative method

$$T_D = \frac{(1-x_1^2 y_1^2)\tau^2}{1-2x_1 y_1 \tau \cos\phi(\frac{1+\tau^2}{2})+\tau^2 x_1^2 y_1^2 + x_1^2 y_1^2 (\sin^2(\phi)(1-\tau^2)^2)-(1-\tau^2)} \frac{n_g L}{c}$$

$T_D \rangle 0$

$T_D = 0$

Slow Light

Normal Light

Figure 7.1 Flowchart for research methodology.

have shown that the deployment of high-bit-rate multi wavelength systems together with optical amplifiers create major nonlinear effects such as SBS, SPM. Nonlinearities stem from the refractive index and attenuation in the intensity of light. The nonlinear effect in optical waveguides is due to the inelastic interaction of photon with material. For nonlinear waveguide, the input threshold power is 0.1 W. If the input power be larger than threshold value, the nonlinear effect appears. If the input power is less, the nonlinear effect can be neglected. In this study, the input power below and above threshold value is used to investigate the behavior of light pulse in term of linearity and nonlinearity.

7.2 DISPERSION IN OPTICAL WAVEGUIDE

A crucial observation is that the physics behind fast light is identical to the physics behind slow light. Although most of us readily accept the notion of a pulse of light moving through a dispersive material at a group velocity less than c, many of us are uncomfortable with the fast light case. Both arise from the same effect. The shifting of the point of constructive interference is to another point in space time. To understand the remarkable slow and fast light properties of pulse propagation in micro-ring resonator systems, the Kramers-Kronig relation connects the real and the imaginary parts of complex response functions of physical systems (Kronig, 1926). In this study three materials have been considered as waveguides. Dispersion of waveguide is examined by this method for generation of fast and slow light. The solutions of the wave equation, in this manner, the refractive index is modified and simulated by Equations (7.1) to (7.4).

$$n = \sqrt{1 + 4\pi \chi} \tag{7.1}$$

where χ is the susceptibility.

The refractive index $n = n' + in''$ can be expressed as:

$$n \cong 1 + 2\pi \chi \tag{7.2}$$

The real and imaginary parts are given by

$$n' = 1 + \delta_{\max} \frac{2(\omega_0 - \omega)\gamma}{(\omega_0 - \omega)^2 + \gamma^2} \tag{7.3}$$

$$n'' = \delta_{\max} \frac{\gamma^2}{(\omega_0 - \omega)^2 + \gamma^2} \tag{7.4}$$

For a near resonant light field, the transmission frequency is denoted by ω_0, 2γ is the width (FWHM) of the resonance. $\delta_{\max}$ is the maximum deviation of the phase index. The maximum derivation of phase index is expressed in Equation (7.5).

$$\delta_{\max} = \frac{\pi N e^2}{2m\omega_0 \gamma} \tag{7.5}$$

Here, N and e are density and charge of electron respectively. The optical waveguide can be used efficiently for the group velocity alteration. The great advantage of waveguide is that they can act as slow or fast light medium directly. Thus, they are benefit into optical information and communication systems. Furthermore, there are many inexpensive and reliable fiber types with different properties available that makes them very flexible to apply. Another important feature is that the slow and fast light effect occurs within dispersive optical waveguide which are used in optical communications.

7.3 SLOW LIGHT GENERATION USING NONLINEAR WAVEGUIDE

Various techniques have been used for the analysis of ring resonators. There are two main classes of such techniques. The first class is an analytical method including the scattering matrix method. The matrix is symmetric because the networks under consideration are reciprocal. The second class is using a graphical approach, it is called the Signal Flow Graph method (SFG) proposed by Mason. This method is originally used in the electrical circuits, which is not widely used in the analysis of optical circuits. In this research scattering matrix method has been used. For reducing the group velocity and finding delay time of output pulse a series of micro-ring resonators consist of three ring resonators coupled to three add drop systems have been used. In this method linear and nonlinear effects for single ring and linear effect for add drop filter have been considered. The proposed system is shown in Figure 7.2.

In this case, bright soliton and Gaussian beam have been used as an input pulse inside to micro-ring resonators. The output of first single ring resonator is given as (Amiri & Ali, 2013e; Ayodeji *et al.*, 2014):

$$E_{out1} = E_{in}y_1x_1 + j\sqrt{\kappa_1}x_1\tau_1 \exp(-j\phi_1)\left(\frac{j\sqrt{\kappa_1}x_1E_{in}}{1 - x_1y_1\tau_1 \exp(-j\phi_1)}\right) \tag{7.6}$$

The throughputs of the first add a drop ring resonator is calculated as follows:

$$E_{t1} = \left(\frac{E_{out1}x_2y_2 - E_{out1}x_2^2x_3y_3 \exp(-\alpha L_2/2) \exp(-j\phi_2)}{1 - x_2y_2x_3y_3 \exp(-\alpha L_2/2) \exp(-j\phi_2)}\right) \tag{7.7}$$

The output of the second ring is:

$$E_{out2} = E_{t1}y_4x_4 + j\sqrt{\kappa_4}x_4\tau_3 \exp(-j\phi_3)\left(\frac{j\sqrt{\kappa_4}x_4E_{t1}}{1 - x_4y_4\tau_3 \exp(-j\phi_3)}\right) \tag{7.8}$$

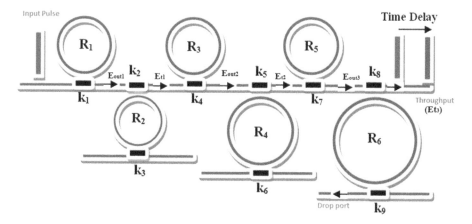

Figure 7.2 Schematic of micro-ring resonator for slow light generation with delay time.

The throughput field of second add drop filter is:

$$E_{t2} = \left(\frac{E_{out2} x_5 y_5 - E_{out1} x_5^2 x_6 y_6 \exp(-\alpha L_4/2) \exp(-j\phi_4)}{1 - x_5 y_5 x_6 y_6 \exp(-\alpha L_4/2) \exp(-j\phi_4)} \right) \qquad (7.9)$$

The output field of third ring (Amiri & Afroozeh, 2014) is:

$$E_{out3} = E_{t2} y_7 x_7 + j\sqrt{\kappa_7} x_7 \tau_5 \exp(-j\phi_5) \left(\frac{j\sqrt{\kappa_7} x_7 E_{t2}}{1 - x_7 y_7 \tau_5 \exp(-j\phi_5)} \right) \qquad (7.10)$$

The throughput field of the third add drop filter is:

$$E_{t3} = \left(\frac{E_{out3} x_8 y_8 - E_{out3} x_8^2 x_9 y_9 \exp(-\alpha L_6/2) \exp(-j\phi_6)}{1 - x_8 y_8 x_9 y_9 \exp(-\alpha L_6/2) \exp(-j\phi_6)} \right) \qquad (7.11)$$

And the output power can then be expressed as:

$$P_{t3} \propto (E_{t3}) \cdot (E_{t3})^* = |E_{t3}|^2 \qquad (7.12)$$

The output signal parameters such as group velocity, FWHM, FSR and delay time from the system are simulated using the MATLAB programming by iterative and numerical methods. All of the other parameters of the ring resonators have been reported in Table 7.1. Generally the delay time can be calculated by considering the transmission in the single ring resonator system using Equations (7.13) to (7.18).

$$T = \frac{E_{out1}}{E_{in}} = \left(\frac{x_1 y_1 - x_1^2 \tau \exp(-j\phi)}{1 - x_1 y_1 \tau \exp(-j\phi)} \right) \qquad (7.13)$$

The external phase shift of MRRs can be achieved from the argument on ratio output field and input field (Heebner et al., 2004) as:

$$\Phi = \arg\left(\frac{E_{out}}{E_{in}} \right) = -i \log\left(\frac{T}{|T|} \right) \qquad (7.14)$$

To achieve the delay time of pulse as a slow light, the external phase shift must be positive. Therefore the term of logarithm in Equation (7.14) should be negative and ratio between transmission and absolute transmission must obey Equation (7.15)

$$0 \langle \frac{\frac{x_1 y_1 - x_1^2 \tau \exp(-j\phi)}{1 - x_1 y_1 \tau \exp(-j\phi)}}{\sqrt{\frac{x_1 y_1 - x_1^2 \tau \exp(-j\phi)}{1 - x_1 y_1 \tau \exp(-j\phi)} \times \frac{x_1 y_1 - x_1^2 \tau \exp(j\phi)}{1 - x_1 y_1 \tau \exp(j\phi)}}} \langle 1 \qquad (7.15)$$

Equation (4.15) shows the first boundary condition which determine rate of radius of ring to generate slow light. This equation gives boundary condition to achieve slow light for single ring resonator. By solving this equation the rate of radius for each waveguide can be calculated. Here, input power, coupler loss and coupler coefficient are fix parameters based on experimental and theoretical previous work and research.

Table 7.2 Available Ring radii to generate slow light in frame of nonlinear waveguide.

Waveguides	InGaAsP/InP	GaAlAs/GaAs	a-Si:H	Selection Radius
Bright Soliton (P = 8W, k = 0.99)	R) 26.16 μm	R) 25.93 μm	R) 25.10 μm	R = 28 μm
Gaussian Pulse (P = 5W, k = 0.99)	R) 26.161 μm	R) 25.92 μm	R) 25.09 μm	R = 28 μm

Therefore, if the input power, coupler coefficient and core area are considered as fix parameters, the regime of radius of ring can be calculated. Because Equation (7.15) depends to coupler coefficient, loss waveguide and internal phase which depends on input power, linear refractive and nonlinear refractive index. Following equation shows the relation between internal phase and input power.

$$\phi = \phi_L + \phi_{NL} = k n_0 L + k L n_2 P / A_{\text{eff}} \tag{7.16}$$

where, the first part and second part are linear and nonlinear phase respectively. P is input power and A_{eff} is the core area.

Based on the Equation (7.15), if using fix parameters such as $k = 0.99$ (Suchat *et al.*, 2010), $P = 8$ W and $A_{\text{eff}} = 0.1$ μm for bright soliton and input the Equation (7.16), the range of radii of ring for slow light generation can be calculated and shown in Table 7.2. For Gaussian beam the input is considered $P = 5$ W, $k = 0.99$ and $A_{\text{eff}} = 0.1$μm Therefore from Equation (7.15) and the characteristic of waveguides, the range of radii can be calculated as given in Table 7.2.

Therefore, in this model under same condition, the radius of ring should be 28 μm and $k = 0.99$ and input power is not more affected to determine of radius. The results show the variation of input power does not influence the ring radius significantly.

Internal phase shift of MRR is given by $(\phi = \omega_0 T_R)$, where ω_0 is one of the resonance frequencies of the resonators and T_R is the transit time of the resonator. The phase sensitivity is obtained by differentiating the external phase shift (Su *et al.*, 2007).

$$\Phi' = \frac{d\Phi}{d\phi}$$

$$= \frac{(1 - x_1^2 y_1^2)\tau^2}{1 - 2x_1 y_1 \tau \cos\phi\left(\frac{1+\tau^2}{2}\right) + \tau^2 x_1^2 y_1^2 + x_1^2 y_1^2(\sin^2(\phi)(1 - \tau^2)^2) - (1 - \tau^2)} \tag{7.17}$$

Therefore the group delay of the ring resonator can be achieved by radian frequency of the transfer function and is defined as:

$$T_D = \frac{d\Phi}{d\omega} = \frac{d\Phi}{d\phi}\frac{d\phi}{d\omega} = \Phi' T_R \tag{7.18}$$

$$T_D = \frac{(1 - x_1^2 y_1^2)\tau^2}{1 - 2x_1 y_1 \tau \cos\phi\left(\frac{1+\tau^2}{2}\right) + \tau^2 x_1^2 y_1^2 + x_1^2 y_1^2(\sin^2(\phi)(1 - \tau^2)^2) - (1 - \tau^2)} \frac{n_g L}{c}$$

$$\tag{7.19}$$

Here, n_g is the group refractive index and L is the circumference of the ring. This equation shows that the group delay achieves its maximum in the resonance wavelength. It is reduced when it detune from resonance. The group index of the materials depends on the dispersion term $\left(\frac{dn}{d\omega}\right)$. If $\frac{dn}{d\omega}\rangle 0$ the dispersion is normal and point of constructive interference occurs at a later time. Therefore slow light is generated. To achieve delay time, T_D should be positive to generate slow light. Thus the Equation (7.19) will be as a boundary condition.

$$\frac{(1 - x_1^2 y_1^2)\tau^2}{1 - 2x_1 y_1 \tau \cos\phi\left(\frac{1+\tau^2}{2}\right) + \tau^2 x_1^2 y_1^2 + x_1^2 y_1^2 (\sin^2(\phi)(1 - \tau^2)^2) - (1 - \tau^2)} \frac{n_g L}{c} \rangle 0 \quad (7.20)$$

Here the delay time as a shift time due to the propagation through the MRRs can be calculated as:

$$\Delta T = N T_R + T_D \quad (7.21)$$

Here, N is number of roundtrip and T_D is delay time. By consider Equation (7.15) and Equation (7.21), T_D must be positive. This relation shows that the group delay is inversely proportional to the group velocity. Therefore, the generation of slow light for optical buffer and read only memory is realized.

The add drop system is used to filter the noise signal and its parameters does not affect the delay time (Amiri, 2011b; Amiri & Ali, 2013c; Amiri et al., 2013a). To retrieve the signal from the chaotic noise, we propose to use the add/drop device with the appropriate parameters. The parameters such as coupler coefficient and ring radius are used to control of output power. In this study, output power is not focused. But these parameters are more important to get desired output power for applications. Thus, we use this system to filter chaotic and noise signals. The optical circuits of ring resonator add/drop filters for throughput and drop port can be given by Equations (7.22) and (7.23).

$$\left|\frac{E_t}{E_{in}}\right|^2 = \frac{(1 - \kappa_1) - 2\sqrt{1 - \kappa_1} \cdot \sqrt{1 - \kappa_2} e^{-\frac{\alpha}{2}L} \cos(k_n L) + (1 - \kappa_2) e^{-\alpha L}}{1 + (1 - \kappa_1)(1 - \kappa_2) e^{-\alpha L} - 2\sqrt{1 - \kappa_1} \cdot \sqrt{1 - \kappa_2} e^{-\frac{\alpha}{2}L} \cos(k_n L)} \quad (7.22)$$

$$\left|\frac{E_d}{E_{in}}\right|^2 = \frac{\kappa_1 \kappa_2 e^{-\frac{\alpha}{2}L}}{1 + (1 - \kappa_1)(1 - \kappa_2) e^{-\alpha L} - 2\sqrt{1 - \kappa_1} \cdot \sqrt{1 - \kappa_2} e^{-\frac{\alpha}{2}L} \cos(k_n L)} \quad (7.23)$$

where E_t and E_d represents the optical fields of the throughput and drop ports respectively. The circumference of the ring is $L = 2\pi R$, here R is the radius of the ring (Kouhnavard et al., 2010a; Amiri & Nikoukar, 2010–2011). The chaotic noise cancellation can be managed by using the specific parameters of the add/drop device, which the required signals can be retrieved by the specific users. K_1 and K_2 are coupling coefficient of add/drop filters, k_n is the wave propagation number in a vacuum, where the waveguide (ring resonator) loss is α. The fractional coupler intensity loss is γ. In the case of add/drop device, the nonlinear refractive index is neglected. The add/drop

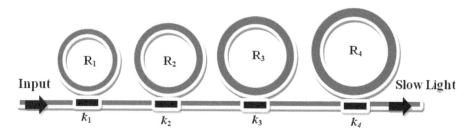

Figure 7.3 Schematic of micro-ring resonator for slow light generation with time delay.

ring resonators are not affect to delay time. Therefore by using proper parameters just help to achieved signals without noise.

In this case, three rings as resonators and three rings as filtering are used. If used more than these rings, the output power decrease and approaches to zero. If reduce the rings, the delay time will be reduce. So the number of rings is importance to improve the delay time.

7.4 SLOW LIGHT GENERATION USING LINEAR WAVEGUIDE

Here, to generate slow light, series of ring resonators is used. In this method, the nonlinear effect of waveguide is neglected because of threshold power of nonlinear effect. Figure 7.3 shows the schematic of proposed system consists of series ring resonators. Bright soliton and Gaussian beam can be used for three waveguide as input pulses. In this method chaotic signal can be canceled by neglecting nonlinear effects. So, in this manner add drop filter is not necessary to filtering chaotic signal. Therefore the determination of phase shift is caused by observed time delay. Time delay can be obtained using phase shift and it shows the rate of slow light in this method. The mathematical analysis of ring is mentioned in last section as a single ring resonator.

Figure 7.4 shows the flowchart that demonstrates the process of slow light generation using micro-ring resonators. This flowchart shows the manner of slow light generation via variable and fixed parameters. In this method bright and Gaussian pulse has been used as input pulse. Linear and nonlinear effects have been considered.

Here, the Equation (7.15) as a boundary condition is used to develop ring configuration. The radii of the ring depend on coupler coefficient. Here the input power and core area are not affective, because the term of nonlinear is neglected and this part depends on input power and core area based on Equation (7.16). Thus in this case the coupler coefficient is affective. By input coupler coefficient as fix parameters, the range of radii of ring resonators can be calculated. Table 7.3 shows the determined ring radii of proposed system and configuration.

In this case, four rings are used as resonators. If more than these rings are used, the output power decreases and approaches to zero. If the rings are reduced, the delay time will be reduced.

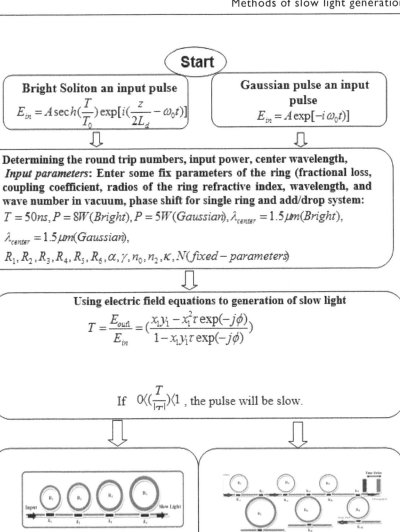

Start

Bright Soliton an input pulse

$$E_{in} = A \operatorname{sech}(\frac{T}{T_0}) \exp[i(\frac{z}{2L_d} - \omega_0 t)]$$

Gaussian pulse an input pulse

$$E_{in} = A \exp[-i\,\omega_0 t)]$$

Determining the round trip numbers, input power, center wavelength, *Input parameters*: Enter some fix parameters of the ring (fractional loss, coupling coefficient, radios of the ring refractive index, wavelength, and wave number in vacuum, phase shift for single ring and add/drop system:

$T = 50ns, P = 8W (Bright), P = 5W (Gaussian), \lambda_{center} = 1.5 \mu m (Bright),$

$\lambda_{center} = 1.5 \mu m (Gaussian),$

$R_1, R_2, R_3, R_4, R_5, R_6, \alpha, \gamma, n_0, n_2, \kappa, N (fixed-parameters)$

Using electric field equations to generation of slow light

$$T = \frac{E_{out1}}{E_{in}} = (\frac{x_1 y_1 - x_1^2 \tau \exp(-j\phi)}{1 - x_1 y_1 \tau \exp(-j\phi)})$$

If $0 < (\frac{T}{|T|}) < 1$, the pulse will be slow.

Slow light in frame of linear

Slow light in frame of nonlinear

Plotting the output, through port power, delay time as a slow light

End

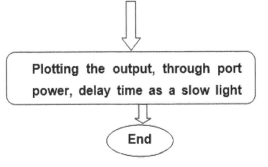

Figure 7.4 Flowchart for slow light generation.

Table 7.3 Ring radii to generate slow light in frame of linear waveguide.

Waveguides	InGaAsP/InP	GaAlAs/GaAs	a-Si:H	Selection Radii
$k_1 = 0.82$	R⟩ 9.71 μm	R⟩ 9.61 μm	R⟩ 9.35 μm	$R = 10$ μm
$k_2 = 0.92$	R⟩ 14.32 μm	R⟩ 14.14 μm	R⟩ 13.75 μm	$R = 15$ μm
$k_3 = 0.97$	R⟩ 19.90 μm	R⟩ 19.73 μm	R⟩ 19.10 μm	$R = 20$ μm
$k_4 = 0.98$	R⟩ 22.21 μm	R⟩ 22.02 μm	R⟩ 19.50 μm	$R = 25$ μm

7.5 EXAMINE OF DISPERSION WAVEGUIDES

All optical waveguide materials to generate slow and fast light should be dispersive. This means that the refractive index varies with wavelength. There are several ways to measure dispersion in materials. A simple measure is the Abbe number (V_D). Another measure of dispersion is the derivative $dn/d\lambda$. Besides absorption resonances, amplification processes are also applicable to induce a material dispersion. Hence, optical waveguide can be used for slow and fast light generation. Dispersion is most often described for light waves, but it may occur for any kind of wave that interacts with a medium or passes through an inhomogeneous geometry. In this study, three waveguide have been used as dispersion material.

They have capability of high bandwidth while the maximum achievable time delay is small compared to other mechanisms. The classical phase shift is determined solely by the angular velocity, the optical frequency, and the area is near resonance and completely independent of the medium's other properties such as the index of refraction and its dispersion properties. The refractive index as a real part versus wavelength is simulated using Kramers-Kronig relation method for three materials which has shown in Figure 7.5. These results show the variation of refractive index is maximized in near resonance based on the physical materials as waveguides. The a-S:H waveguide have the maximum variation refractive index.

Figure 7.6 demonstrates negative and positive group refractive index. It shows the variation of group refractive index versus wavelength. It illustrates in the positive area in shoulder of the curve, the light can be slow down and in the negative area the light can be fast up. It shows that the most variation of refractive index occurs in near resonance. When light travels in negative index then it can be fast up, and at the positive index can be slow down. According to Figures 7.5 and 7.6, frequency region close to resonance wavelength can be used to generate slow light.

The scattering process only occurs over a narrow range of frequencies, which means that the control beam creates a resonant region in which the response of the waveguide to light is maximal. This resonance has a width of approximately 10 MHz in a standard telecommunications optical waveguide. However, a big advantage of this approach is that the central frequency of the resonance that is responsible for the slow and fast light effect can be changed by simply changing the frequency of the control beam. The adjustable behavior of the group velocity can be used to engineer systems with large externally controllable dispersion, where $\frac{dn}{d\omega}$ have very large positive or negative values. Therefore it was possible to propagate pulses faster or slower. In a spectral area of normal dispersion $\left(\frac{dn}{d\omega}\rangle 0\right)$ the group velocity decreases. In the

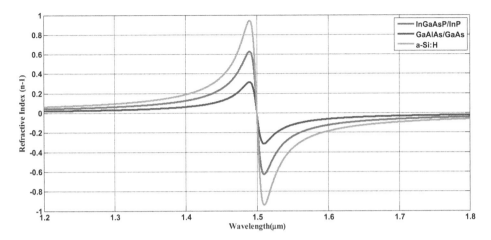

Figure 7.5 The variation of refractive index versus wavelength in three waveguide.

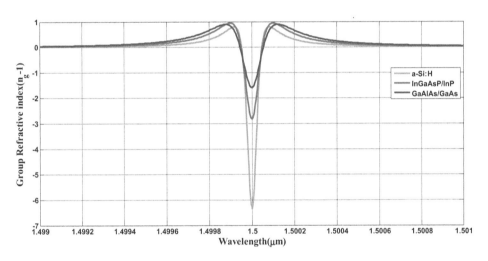

Figure 7.6 Schematic of variation group index in three waveguide against wavelength.

waveguide each wave of different frequency propagates with different phase velocity. On the shoulder of the curve, the group index is more than one. Therefore in this area, the pulse propagation becomes slow down. Near the resonance the slope of the curve becomes very steep indicating that the phase is sensitive. Figure 7.7 shows the variation of the dispersion versus wavelength.

Near each resonance the phase shift varies rapidly with the frequency leading to reduce the group velocity Therefore in the range of wavelengths which are close to resonance, the light can be fast up and in the shoulder the light can be slow down. Figure 7.8 shows the variation of group velocity versus waveguides. It shows that the

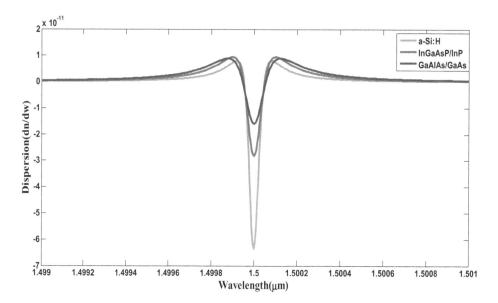

Figure 7.7 Variation of dispersion in waveguide versus wavelength.

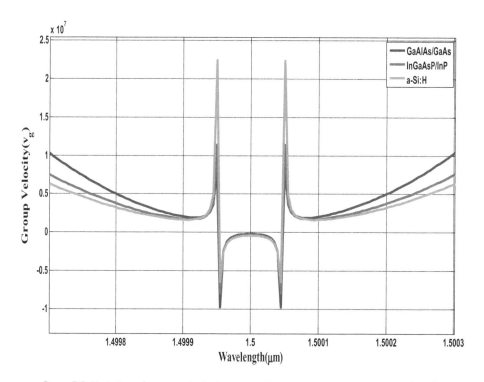

Figure 7.8 Variation of group velocity in waveguide near resonance versus wavelength.

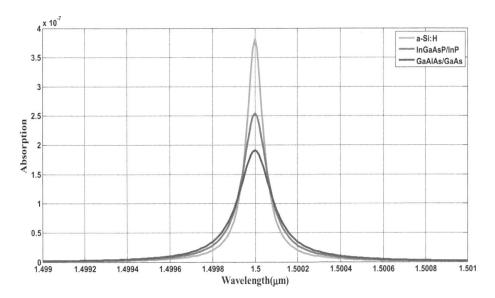

Figure 7.9 The variation of absorption in three waveguide versus wavelength.

fast light can be achieved near resonance wavelength. This effect appears in a range of wavelengths which are close to the resonance wavelengths of the material.

The variation linear absorption coefficient (α) can be obtained and simulated is shown in Figure 7.9. In the wavelength resonance, the absorption is the maximum. Large absorption coefficient occurs at near resonance frequencies where a-Si: H is the largest absorption coefficient in these materials. The results show that the best center wavelength to use in input pulse for these waveguides is 1.5 μm.

7.6 APPLICATIONS

Controlling the group velocity of light pulses is useful to achieve various functionalities in optical communication and network. The main purpose of proposed system is to manage large volumes of data that it receives and send it to the right destination. Storing of data is commonly referred to as buffering. The stored data can be retrieved at will and sent to the right destination when the data traffic in that particular channel clears up. The selected light pulse can be trapped and used to perform the memory which is controlled by light. The adiabatic storing pulse process to preserve the coherent information encoded can also be performed. The key advantages of the system are the reversely compress bandwidth and the maintaining power, which can be tuned to obtain the arbitrary pulse for Read Optical Memory (ROM).

Chapter 8

Soliton generation and transmission in optical fiber link

8.1 SOLITON CHAOTIC SIGNAL GENERATION USING THE MRRS

In the case of single ring resonator (R_1) and an add/drop filter system (R_{ad}), the used parameters are shown in Table 8.1.

The nonlinear behaviors of the fiber optic ring resonator in 20,000 round-trips inside the optical fiber ring resonator was described by Amiri *et al.* The input power is maximized to 1 W, is inserted to a single MRR system, where the output power is varied directly with the input power. The output electrical power will be reduced as well as improved beyond the particular input electrical power abruptly, giving the particular output power having a couple values named the bistability characteristics, that is certainly switched-on and switched-off. The output powers at the round-trips 5750 times has shown the characteristics called "bifurcation". At this point, the abrupt change within the input electrical power provides output electrical power along with a pair of values. This is known as the optical bistability, the spot that the optical power switched-on/off happen. The bifurcation behavior occurs ahead of the chaotic signal. The chaotic signal can be generated and controlled by varying the coupling coefficients, where the required output power is obtained.

Figure 8.1 shows the chaotic signals generation for variety of coupling coefficients, where Figure 8.1(a–b) shows the output signals of the single ring resonator in terms

Table 8.1 Parameters of the system.

Parameters	Value
R_1 = radius of the ring	15 μm
κ = coupling coefficient of the ring	0.0225
R_{ad} = Add/drop MRR system, radius	15 μm
κ_1 = coupling coefficient of the add/drop	0.01
κ_2 = coupling coefficient of the add/drop	0.01
λ_0 = central wavelengths of the Gaussian laser	1.55 μm
A_{eff} = effective mode core area	0.30 μm^2
α = waveguide (ring resonator) loss	0.02 dB km^{-1}
γ = fractional coupler intensity loss	0.01
n_0 = linear refractive index	3.34 (InGaAsP/InP)
n_2 = nonlinear refractive index	3.8×10^{-20} m^2/W

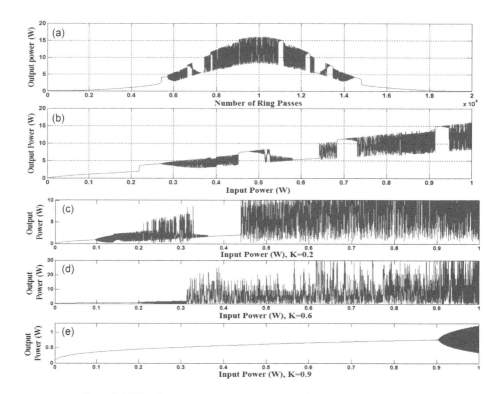

Figure 8.1 The chaotic signal generation within the single ring resonator.

of round-trips and input power. Figure 8.1(c–e) show the output signals for different coupling coefficients of $\kappa = 0.2$, 0.6 and 0.9 respectively. Within practical applications, the input power is required to become lower as a result of available industrial laser diodes. As a result, a MRR could present the actual chaotic behavior using reduced input electric power, that is suited to assistant carry out to the communication system as well as device manufacture.

8.2 SINGLE DARK AND BRIGHT SOLITON GENERATION

To recover the pulses from the chaotic noises in the fiber ring resonator, the use of an add/drop filter system with the appropriate parameters is recommended. Figure 8.2 shows the output signals of the add/drop filter system, where the single soliton pulses of dark and bright can be obtained. Here the temporal form of these signals is presented.

The attenuation, or loss in signal power, resulting from the insertion of a component, such as a coupler or splice, in a circuit. Insertion loss is measured as a comparison of signal power at the point the incident energy strikes the component and the signal power at the point it exits the component. Insertion loss typically is measured in decibels (dB), although it also may be expressed as a coefficient or a fraction. The insertion loss of the add/drop filter system is show in Figure 8.3 which shows that how the bandwidth of the generated single pulse can be controlled via the system.

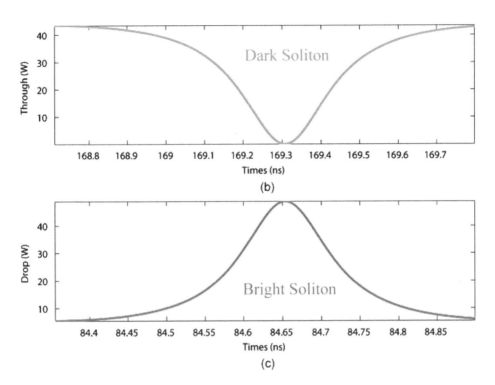

Figure 8.2 Temporal dark and bright signals using the add/drop filter system.

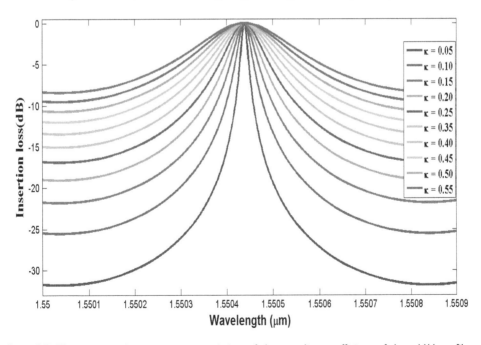

Figure 8.3 The insertion loss, respect to variation of the coupling coefficient of the add/drop filter system.

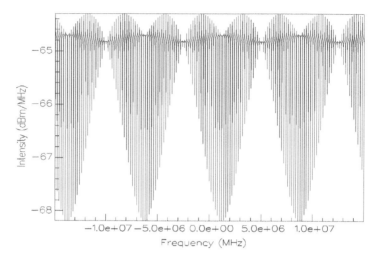

Figure 8.4 Throughput output signals of the add/drop ring resonator with $L = 750\,\mu m$, $\kappa_1 = \kappa_2 = 0.02$, $\alpha = 0$.

Therefore, the bandwidth varies respect to the variation of the coupling coefficients of the add/drop filter system. Here the increase of the coupling coefficient leads to increase the bandwidth as it can be seen from Figure 8.3.

8.3   SOLITON COMB GENERATION USING THE ADD/DROP SYSTEM

The throughput output signals of the add/drop filter system with two waveguide and coupling factor of $\kappa_1 = \kappa_2 = 0.02$ in both symmetrical couplers is shown in Figure 8.4.

The output intensity at the drop port will is shown in Figure 8.5, which indicates that the resonance wavelength is fully extracted by the resonator when $\kappa_1 = \kappa_2$ and $\alpha = 0$.

The group delay profile of the drop port output referenced to the input port is shown in Figure 8.6.

8.4   ADD/DROP FILTER SYSTEM INCORPORATING WITH SERIES OF RING RESONATOR

Exciting new technological progress, particularly in the field of the MRR interferometers, provide the foundation for the development of new transmission techniques. The highly chaotic signals can be generated and sliced into ultra-short single and multi-soliton pulses (Ali *et al.*, 2010u; Amiri *et al.*, 2012b; Alavi *et al.*, 2014). The storage of optical soliton pulses in picometer and femtoseconds can be performed using the proposed system, where the multi-soliton generation is the advantage for the systems

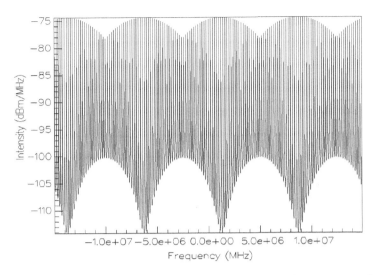

Figure 8.5 Drop port output of an add/drop ring resonator with $R = 750 \, \mu m, \kappa_1 = \kappa_2 = 0.02, \alpha = 0$.

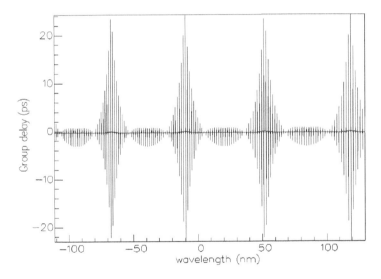

Figure 8.6 Group delay of the add-drop ring resonator with $R = 750 \, \mu m, \kappa_1 = \kappa_2 = 0.02, \alpha = 0$.

of ring resonators. Add-drop filters using MRR have shown great promise for practical applications as such resonators have very high quality factors (up to 10^9). MRRs have been studied most thoroughly due to their ease of fabrication and on chip structure.

Wavelength selective optical add-drop system is required based on the optical access networks. Add-drop system used in communication should have a good reflection characteristic, a narrow spectral bandwidth, and a low implementation cost. For those reasons, many researchers have been proposed various technologies for

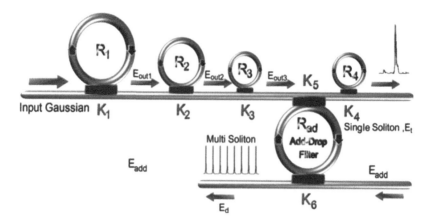

Figure 8.7 A schematic of the proposed MRR's system, where R_s: ring radii, κ_s: coupling coefficients, R_d: an add-drop ring radius.

implementation of the add-drop filter (Saktioto *et al.*, 2010c; Amiri & Ali, 2013; Amiri & Afroozeh, 2014d). The performance of various add-drop multiplexers are compared based on efficiency, number of tributaries and speed limitations. Depending on the wavelength of the signal and also having a wavelength "add" function in which optical signals presented to the add port(s) are also transferred to the output port. The system of the ring resonator interferometer is shown in Figure 8.7.

The tunable narrow band laser system such as the MRR interferometers, provide the basis for the development of new transmission techniques. The system of the ring resonator interferometer can be used to generate narrow bandwidth. A series of MRRs connected to an add-drop interferometer system is proposed. To control the output signals from the through and drop ports of the add-drop filter system, an additional input signal can be added to the add port of the system. The nonlinear refractive index of the micro-ring is $n_2 = 3.4 \times 10^{-17}$ m^2/W. The capacity of the output signals can be improved through the generation of peaks with smaller FWHMs, thus in particular, adiabatically perturbed picometer and femtosecond solitons in single-mode fibres are revealed. To transmit the soliton signals via long distance communications, ultra-short soliton pulses are required. Here we characterize generation of femtoseconds optical soliton pulses with Full Width at Half-Maximum (FWHM) smaller than 100 fs with respect to the ring's radius variation of the system. As a result, localized single spatial and temporal soliton pulses can be generated to form high capacity and secure transmission signals, applicable in optical soliton communications.

The input optical fields (E_{in}) in the form of Gaussian pulse can be expressed by (Afroozeh *et al.*, 2011b; Amiri *et al.*, 2013c)

$$E_{in}(z, t) = E_0 \exp\left[\left(\frac{z}{2L_D}\right) - i\omega_0 t\right] \tag{8.1}$$

Here E_0 and z are the optical field amplitude and propagation distance, respectively. T represents soliton pulse propagation time in a frame moving at the group velocity, $(T = t - \beta_1 \times z)$, where β_1 and β_2 are the coefficients of the linear and second order terms of the Taylor expansion of the propagation constant. The dispersion length of the soliton pulse can be defined as $L_D = T_0^2/|\beta_2|$, where the frequency carrier of the soliton is ω_0. The intensity of soliton peak is $(|\beta_2/\Gamma T_0^2|)$, where T_0 is representing the initial soliton pulse propagation time. A balance should be achieved between the dispersion length (L_D) and the nonlinear length $(L_{NL} = 1/\Gamma\phi_{NL})$, where $\Gamma = n_2 \times k_0$, is the length scale over which disperse or nonlinear effects causes the beam becomes wider or narrower. Here, $L_D = L_{NL}$. The total index (n) of the system is given by (Amiri et al., 2014e; Amiri et al., 2014d).

$$n = n_0 + n_2 I = n_0 + \left(\frac{n_2}{A_{\text{eff}}}\right) P, \tag{8.2}$$

where n_0 and n_2 are the linear and nonlinear refractive indices, respectively. I and P are the optical intensity and optical power, respectively. A_{eff} represents the effective mode core area of the device, where in the case of MRRs, the effective mode core areas range from 0.50 to 0.1 μm^2. The normalized output of the light field is defined as (Ali et al., 2010n; Amiri et al., 2012e; Amiri et al., 2014c).

$$\left|\frac{E_{\text{out}}(t)}{E_{\text{in}}(t)}\right|^2 = (1 - \gamma) \times \left[1 - \frac{(1 - (1 - \gamma)x^2)\kappa}{(1 - x\sqrt{1-\gamma}\sqrt{1-\kappa})^2 + 4x\sqrt{1-\gamma}\sqrt{1-\kappa}\sin^2\left(\frac{\phi}{2}\right)}\right] \tag{8.3}$$

Here, κ is the coupling coefficient, $x = \exp(-\alpha L/2)$ represents a round-trip loss coefficient, $\phi_0 = kLn_0$ and $\phi_{NL} = kLn_2|E_{\text{in}}|^2$ are the linear and nonlinear phase shifts and $k = 2\pi/\lambda$ is the wave propagation number and γ is the fractional coupler intensity loss. Here L and α are the waveguide length and linear absorption coefficient, respectively. The input power insert into the input port of the add-drop interferometer system. E_{th} and E_{drop} represent the optical electric fields of the through and drop ports, respectively, therefore,

$$E_{\text{th}} = \frac{-E_{\text{out 3}} \times \kappa_5 \times \sqrt{1-\kappa_6} \times e^{\frac{-\alpha L_{\text{ad}}}{2}-jk_n L_{\text{ad}}}}{1 - \sqrt{(1-\kappa_5) \times (1-\kappa_6)} \times e^{\frac{-\alpha L_{\text{ad}}}{2}-jk_n L_{\text{ad}}}}$$

$$+ \frac{-E_{\text{add}} \times \sqrt{(\kappa_5 \times \kappa_6) \times (1-\kappa_5) \times (1-\kappa_6)} \times e^{\frac{-3\alpha L_{\text{ad}}}{4}-jk_n \frac{3L_{\text{ad}}}{2}}}{1 - \sqrt{(1-\kappa_5) \times (1-\kappa_6)} \times e^{\frac{-\alpha L_{\text{ad}}}{2}-jk_n L_{\text{ad}}}} \tag{8.4}$$

$$- E_{\text{add}} \times \sqrt{\kappa_5 \times \kappa_6} \times e^{\frac{-\alpha L_{\text{ad}}}{4}-jk_n\frac{L_{\text{ad}}}{2}} + E_{\text{in}} \times \sqrt{1-\kappa_5}$$

$$E_{\text{drop}} = \frac{-E_{\text{out 3}} \times \sqrt{\kappa_5 \times \kappa_6} \times e^{\frac{-\alpha}{2}\frac{L_{\text{ad}}}{2}-jk_n\frac{L_{\text{ad}}}{2}}}{1 - \sqrt{(1-\kappa_5) \times (1-\kappa_6)} \times e^{\frac{-\alpha L_{\text{ad}}}{2}-jk_n L_{\text{ad}}}}$$

$$\tag{8.5}$$

$$- \frac{E_{\text{add}} \times \kappa_6 \times \sqrt{1-\kappa_5} \times e^{\frac{-\alpha}{2}L_{\text{ad}}-jk_n L_{\text{ad}}}}{1 - \sqrt{(1-\kappa_5) \times (1-\kappa_6)} \times e^{\frac{-\alpha L_{\text{ad}}}{2}-jk_n L_{\text{ad}}}} + E_{\text{add}} \times \sqrt{1-\kappa_6}$$

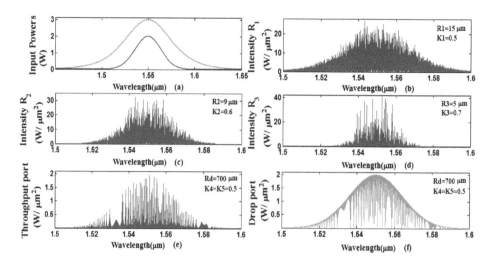

Figure 8.8 Results of spatial dark and bright soliton generation, where (a): inputs bright soliton and Gaussian beam, (b–d): chaotic signals from three rings, (e): bright soliton generation at the through port, (f): dark soliton generation at the drop port.

where $k_n = 2\pi/\lambda$ is the wave propagation number in vacuum and κ_5 and κ_6 are the coupling coefficients, $L_{ad} = 2\pi R_{ad}$ and R_{ad} is the radius of the add-drop interferometer system.

To control the output signals from the through and drop ports of the add-drop filter system, an additional input signal can be added to the add port shown in Figure 8.8. Optical fields of Gaussian pulse with input powers of 3 and 2 W insert into the input and add ports of the system respectively. The radii of the rings are selected to $R_1 = 15\,\mu m$, $R_2 = 9\,\mu m$, $R_3 = 5\,\mu m$, $\kappa_1 = 0.5$, $\kappa_2 = 0.6$, $\kappa_3 = 0.7$, and $A_{eff} = 0.50$, 0.25 and 0.10 μm^2, where the add-drop filter has a radius of $R_d = 700\,\mu m$ and coupling coefficients of $\kappa_5 = \kappa_6 = 0.5$. In this case the fourth ring resonator is ignored. The output signals from the three MRRs and the through and drop ports are shown in Figure 8.8.

The nonlinear refractive index of the micro-ring is $n_2 = 3.4 \times 10^{-17}$ m^2/W. Here dark and bright soliton with FWHM and FSR of 10 pm and 163 pm are simulated as shown in Figure 8.9.

For security purposes, a recommended approach is to use a dark soliton pulse in overcoming power losses, thereby solving the problem of power attenuation (Amiri & Afroozeh, 2014; Amiri *et al.*, 2015c).

The soliton pulses are so stable that the shape and velocity are preserved while travelling along the medium. The capacity of the output signals can be improved through the generation of peaks with smaller FWHMs, thus in particular, adiabatically perturbed picometer, femtosecond solitons in single-mode fibres are revealed. To transmit the soliton signals via long distance communications, ultra-short soliton pulses are required. Therefore, localized single spatial and temporal soliton pulses

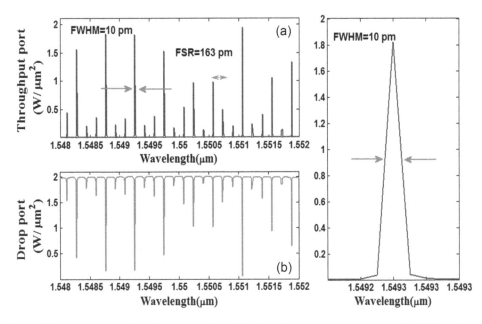

Figure 8.9 Simulation results of spatial dark and bright soliton generation, where (a): bright soliton generation at the through port with FWHM and FSR of 10 pm and 163 pm respectively, (b): dark soliton generation at the drop port with FWHM and FSR of 10 pm and 163 pm respectively.

can be generated to form high capacity and secure transmission signals, applicable in optical soliton communications.

Here we focus on the generation of femtoseconds optical soliton pulses with FWHM smaller than 100 fs respect to the ring's radius variation of the system shown in Figure 1. The nonlinear refractive index is selected to $n_2 = 2.5 \times 10^{-17} \, \mathrm{m^2/W}$, where the coupling coefficients are $\kappa_1 = 0.3$, $\kappa_2 = 0.5$, $\kappa_3 = 0.7$, $\kappa_4 = 0.9$ and $A_{\mathrm{eff}} = 0.50, 0.25, 0.12$ and $0.12 \, \mu\mathrm{m}^2$. Figure 8.10 shows the FWHM of the pulses with regards to the variation of the ring's radius. In order to simulate pulses with FWHM smaller than 100 fs, different orders of the ring's radius have been selected, where the radius of the fourth ring is the variable parameter.

The compressed bandwidth with smaller group velocity is obtained within the ring R_2. The amplifier gain is obtained within the R_3 micro-ring device. The temporal soliton pulse can be formed by using constant gain condition, where a small group velocity is seen. Figure 8.11 shows the results of temporal optical soliton pulses localized within the MRRs with 20,000 round-trip, where an optical ultra-short temporal soliton of FWHM = 83 fs is generated. The input pulse is the Gaussian pulse with power of 500 mW. The nonlinear refractive index is selected to $n_2 = 2.5 \times 10^{-17} \, \mathrm{m^2/W}$. Here, the ring radii are $R_1 = 10 \, \mu\mathrm{m}$, $R_2 = 5 \, \mu\mathrm{m}$, $R_3 = 4 \, \mu\mathrm{m}$, $R_4 = 4 \, \mu\mathrm{m}$ with coupling coefficient of $\kappa_1 = 0.3$, $\kappa_2 = 0.5$, $\kappa_3 = 0.7$, $\kappa_4 = 0.9$.

Generating multi-soliton pulses has become an interesting approach to enlarging communication channel capacity.

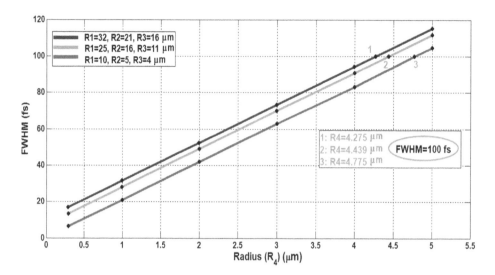

Figure 8.10 Simulation of FWHM versus variable radius of the fourth ring resonator.

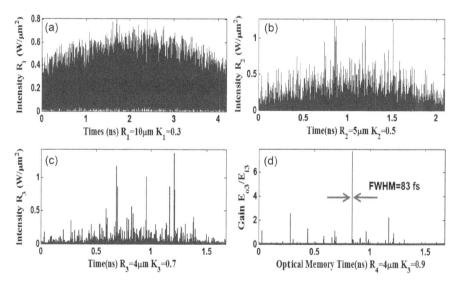

Figure 8.11 Results of temporal soliton generation, where (a): chaotic signals from R_1, (b): chaotic signals from R_2, (c): filtering signals, (d): localized temporal soliton with FWHM of 83 fs.

8.5 RING RESONATOR SYSTEM ANALYSIS TO OPTIMIZE THE SOLITON TRANSMISSION

Using the system consists of three MRRs including R_1, R_2 and R_3, connected to an add/drop filter, the analysis of the soliton signals can be obtained in order to optimize the system. The chaotic signals can be generated within the MRR system when the

Gaussian pulse with input power of 120 mW is inserted into the system. Generation of chaotic signals with respect to the ring's radius has been studied. The coupling coefficient affects the output power significantly, thus in order to generate signals with higher output power, the smaller coupling coefficient can be used. Here the output power of the system is characterized with respect to the different coupling coefficients of the system. The nonlinear refractive index of the MRR is $n_2 = 2.2 \times 10^{-17}\ \text{m}^2/\text{W}$. The capacity of the output signals can be increased through generation of peaks with smaller full width at half maximum (FWHM). Here, we generate and characterize the ultra-short optical soliton pulses respect to the ring's radius and coupling coefficients variation of the system. As a result, soliton pulses with FWHM and free spectral range (FSR) of 50 pm and 1440 pm are generated.

The input optical fields (E_{in}) in the form of Gaussian beam can be expressed by

$$E_{\text{in}}(z,t) = E_0 \exp\left[\left(\frac{z}{2L_D}\right) - i\omega_0 t\right] \tag{8.6}$$

Here E_0 and z are the optical field amplitude and propagation distance, respectively. The dispersion length of the soliton pulse can be defined as $L_D = T_0^2/|\beta_2|$, where T_0 is the propagation time, the frequency carrier of the soliton is ω_0, where the β_2 is the coefficients of the second order terms of the Taylor expansion of the propagation constant. The intensity of soliton peak is ($|\beta_2/\Gamma T_0^2|$), where T_o is representing the initial soliton pulse propagation time (Amiri, 2014). A balance should be achieved between the dispersion length (L_D) and the nonlinear length ($L_{NL} = 1/\Gamma\phi_{NL}$), where $\Gamma = n_2 \times k_0$, is the length scale over which disperse or nonlinear effects causes the beam becomes wider or narrower. Here, $L_D = L_{NL}$. The total index (n) of the system is given by.

$$n = n_0 + n_2 I = n_0 + \left(\frac{n_2}{A_{\text{eff}}}\right)P, \tag{8.7}$$

where n_0 and n_2 are the linear and nonlinear refractive indices respectively. I and P are the optical intensity and optical power, respectively. A_{eff} represents the effective mode core area of the device, where in the case of MRRs, the effective mode core areas range from 0.50 to 0.1 μm^2. The normalized output of the light field is defined (Amiri et al., 2014a; Amiri et al., 2014h) as

$$\left|\frac{E_{\text{out}}(t)}{E_{\text{in}}(t)}\right|^2 = (1-\gamma) \times \left[1 - \frac{(1-(1-\gamma)x^2)\kappa}{(1-x\sqrt{1-\gamma}\sqrt{1-\kappa})^2 + 4x\sqrt{1-\gamma}\sqrt{1-\kappa}\sin^2\left(\frac{\phi}{2}\right)}\right]$$

$$\tag{8.8}$$

Here, κ is the coupling coefficient, $x = \exp(-\alpha L/2)$ represents a round-trip loss coefficient, $\phi = \phi_0 + \phi_{NL}$, $\phi_0 = kLn_0$ and $\phi_{NL} = kLn_2|E_{\text{in}}|^2$ are the linear and nonlinear phase shifts and $k = 2\pi/\lambda$ is the wave propagation number and γ is the fractional coupler intensity loss . Here L and α are the waveguide length and linear absorption coefficient, respectively. The input power insert into the input port of the add/drop filter system. E_{th} and E_{drop} represent the optical electric fields of the through and drop ports, respectively expressed by Equations (8.9) and (8.10),

$$|E_t|^2/|E_{out3}|^2 = \frac{(1 - \kappa_{41}) - 2\sqrt{1 - \kappa_{41}} \cdot \sqrt{1 - \kappa_{42}}e^{-\frac{\alpha}{2}L_d}\cos(k_n L_d) + (1 - \kappa_{42})e^{-\alpha L_d}}{1 + (1 - \kappa_{41})(1 - \kappa_{42})e^{-\alpha L_d} - 2\sqrt{1 - \kappa_{41}} \cdot \sqrt{1 - \kappa_{42}}e^{-\frac{\alpha}{2}L_d}\cos(k_n L_d)}$$

(8.9)

$$|E_d|^2/|E_{out3}|^2 = \frac{\kappa_{41}\kappa_{42}e^{-\frac{\alpha}{2}L_d}}{1 + (1 - \kappa_{41})(1 - \kappa_{42})e^{-\alpha L_d} - 2\sqrt{1 - \kappa_{41}} \cdot \sqrt{1 - \kappa_{42}}e^{-\frac{\alpha}{2}L_d}\cos(k_n L_d)}$$

(8.10)

where $|E_t|^2$ and $|E_d|^2$ are the output intensities of the through and drop ports respectively (Alavi *et al.*, 2013a; Soltanian & Amiri, 2014).

For the first single ring resonator, the parameters were fixed to $\lambda_0 = 1.55\,\mu m$, $n_0 = 3.34$, $A_{eff} = 30\,\mu m^2$, $\alpha = 0.01\,dB\,mm^{-1}$, and $\gamma = 0.1$. The length of the ring has been selected to $L = 60\,\mu m$, where the coupling coefficient is fixed to $\kappa = 0.0225$ and the linear phase shift has been kept to zero. The total round-trip of the input pulse inside the ring system was 20,000. The ring resonator is considered as a passive filter system which can be used to generate signals in the form of chaos, applied in optical communication with regards to suitable parameter of the system. Figure 8.12 shows that bifurcation and chaotic behaviour of the single ring resonator system, where the Gaussian beam with input power of 120 mW is used.

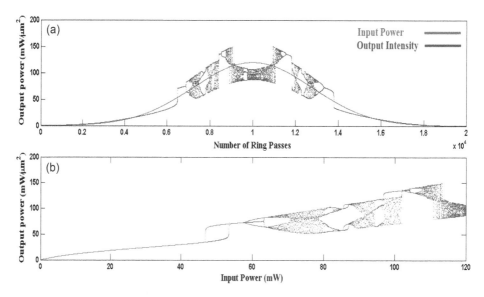

Figure 8.12 Bifurcation and chaos in single ring resonator with $L = 60\,\mu m$, where (a): Output intensity (mW/μm^2) versus round-trip, (b): Output intensity (mW/μm^2) versus input power (mW).

Beside the ring resonator' radius parameter, the coupling coefficient of the single ring resonator also is considered to be an effective parameter to determine the output intensity power of the system. In order to characterize the ring resonator system based on the coupling coefficient, the circumference of the ring has been selected to $L = 40\,\mu$m to avoid the chaotic signals. Therefore, chaotic signals are neglected. Figure 8.13 shows the dependence of the output power of the ring resonator system on the coupling coefficient.

Therefore, output power of the system decreases with increase the coupling coefficient as shown in Figure 8.13. In order to optimize the system, the smaller coupling coefficient is recommended.

In Figure 8.14(a), the input Gaussian beam has $50\,$ns pulse width and peak power of 2 W. The ring radii are $R_1 = 15\,\mu$m, $R_2 = 9\,\mu$m, $R_3 = 7\,\mu$m, $R_d = 80\,\mu$m and $\kappa_1 = 0.96$, $\kappa_2 = 0.94$, $\kappa_3 = 0.92$, $\kappa_4 = \kappa_5 = 0.1$. The fixed parameters are selected to $\lambda_0 = 1.55\,\mu$m, $n_0 = 3.34$ (InGaAsP/InP), $A_{\text{eff}} = 0.50, 0.25$ and $0.10\,\mu$m^2, $\alpha = 0.5\,$dBmm^{-1}, $\gamma = 0.1$. The nonlinear refractive index is $n_2 = 2.2 \times 10^{-17}\,$m^2/W. Optical signals are sliced into smaller signals broadening over the band as shown in Figures 8.14 (b–d). Therefore, large bandwidth signal is formed within the first ring device, where compress bandwidth with smaller group velocity is attained inside the ring R_2 and R_3, such as filtering signals. Localized soliton pulses are formed within the add/drop filter system, where resonant condition is performed, given in Figures 8.14 (e–h). However, there are two types of dark and bright soliton pulses. Here the multi bright soliton pulses with FSR and FHWM of 1440 pm, and 50 pm are simulated.

The variation of the FWHM versus the coupling coefficients (κ_1) and (κ_2) is shown in Figure 8.15. Thus increasing the coupling coefficients leads to increase the FWHM. The variation of the FWHM versus the coupling coefficients (κ_3) and radius of the

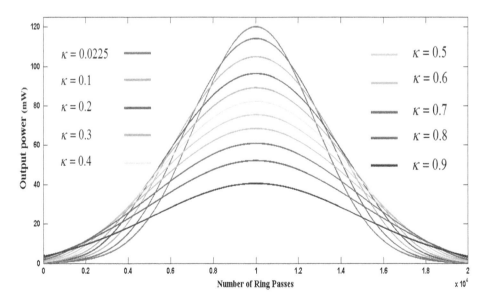

Figure 8.13 The output power of the ring resonator versus round-trip with respect to different coupling coefficients used.

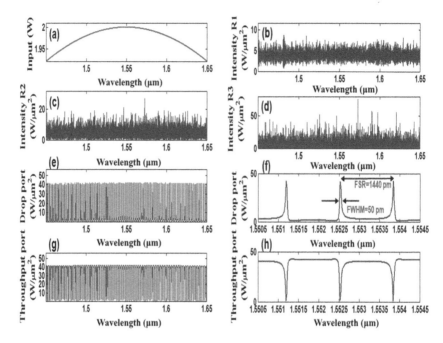

Figure 8.14   Results of the multi-soliton pulse generation, (a): input Gaussian beam, (b–d): large band-width signals, (e–f): bright soliton with FSR of 1440 pm, and FWHM of 50 pm, (g–h): dark soliton with FSR of 1440 pm, and FWHM of 50 pm.

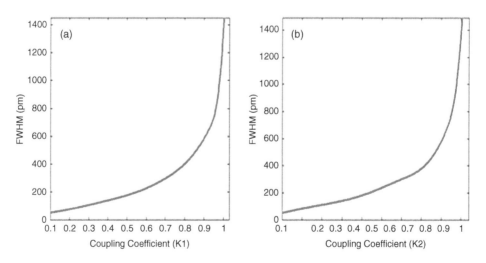

Figure 8.15   Simulation of FWHM, where (a): coupling coefficient (κ_1) of the first ring varies, (b): coupling coefficient (κ_2) of the second ring varies.

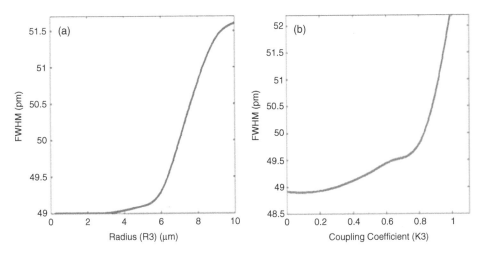

Figure 8.16 Simulation of FWHM, where (a): radius of the third ring varies, (b): coupling coefficient of the third ring varies.

third ring resonator is shown in Figure 8.16. Here, the same concept is valid, thus increasing the variable parameters such as the ring radius and coupling coefficient of the three rings connected to the add/drop filter system causes the FWHM increased. To maximize the efficiency of the MRRs, the resonator bandwidth should be selected properly. This still allows for selection between a small, high-finesse resonator or a larger and proportionally lower-finesse resonator.

Therefore, the low finesse is a benefit for an optical transmitter system in which the system experiences uniform transmission along the fiber optics, where the higher finesse shows better performance (sensitivity) of this system. The variation of the FWHM and FSR versus the ring's radius of the add/drop system is shown in Figure 8.17.

Using this method, the output power of the system can be simulated successfully. This system act as a passive filter system which can be used to split the input power and generate chaotic signals using suitable parameters of the system (Ali *et al.*, 2010t; Fazeldehkordi *et al.*, 2014). Therefore input power of Gaussian beam can be sliced to smaller peaks as chaotic signals. The chaotic signals have many applications in optical communications.

Therefore, the dependence of the chaotic signals on the radius of the ring resonator has been investigated. The output power of the system depends on the coupling coefficient, where higher coupling coefficient leads to generate pulses with lower output power, thus the system can be improved using smaller coupling coefficient. The series of MRRs are connected to an add/drop filter system to generate ultra-short soliton pulses. The soliton pulses were generated using the proposed system, where ultra-short soliton pulse with FWHM of 50 pm are obtained and analyzed regarding the variable parameter such as the radius and coupling coefficient of the rings. Optical channel filters with wide FSR (high selectivity) are required in such a system like DWDM in optical communication, where, the low finesse is a benefit for an optical transmitter system in which the system experiences uniform transmission along the fiber optics.

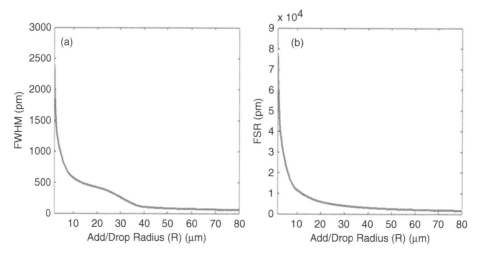

Figure 8.17 FWHM and FSR, where (a): Add/Drop's radius (R) versus FWHM, (b): Add/Drop's radius (R) versus FSR.

8.6 RING RESONATOR FOR COMMUNICATION APPLICATIONS

In recent years a growth in demand for wireless network technologies motivated the development of a new network technology standard (Chang *et al.*, 2012). Since January 2007 IEEE workgroup has been developing IEEE 802.16 standard which in turn developed into the IEEE 802.16m standard or Broadband Wireless Access (BWA) better known as Worldwide Interoperability for Microwave Access (WiMAX) (Nafea *et al.*, 2013). IEEE 802.16m amends the IEEE 802.16e Wireless Metropolitan Area Network (WMAN) standard to meet the cellular layer requirements of the International Telecommunication Union–Radio communication International Mobile Telecommunications (ITUR/IMT)–advanced next generation mobile networks (4G systems) (So-In *et al.*, 2009). The WiMAX air interface promises high bandwidth rates, capable of data transfer rates up to 1 Gb/s depending on the available bandwidth and multiple antenna mode, which can cover metropolitan area of several kilometers (Coexistence, 2009). In theory, a WiMAX can cover an area of 35 miles (56 kms) for fixed stations and 3 to 10 miles (5 to 15 kms) for mobile stations. The foreseen backward compatibility of IEEE 802.16 m standard enables the smooth evolution of current WiMAX systems and easy deployment with the legacy mobile stations (MSs) and base stations (BSs). While the technology mostly is primarily focused on IP-based services, it also supports Ethernet as it is an important factor for some fixed access deployments (Srikanth *et al.*, 2012).

WiMAX air interface includes the Medium Access Control (MAC) and physical (PHY) layers of BWA in which multiple accesses is achieved through Orthogonal Frequency-Division Multiplexing Access (OFDMA) in the PHY layer of the air interface that assigns a subset of subcarriers to each individual user (Lange *et al.*, 2012;

Carro-Lagoa *et al.*, 2013). OFDM enables Downlink (DL) and Uplink (UL) Multi-Input Multi-Output (MIMO) as well as Beam Forming (BF) features. Furthermore, IEEE 802.16m uses multi-hop relay architectures for improved coverage and performance (Chuang *et al.*, 2012). The WiMAX is similar to the wireless standard known as Wi-Fi, but on a much larger scale and at faster speeds. A nomadic version would keep WiMAX-enabled devices connected over large areas, much like today's cell phones. A single WiMAX antenna is expected to have a range of up to 40 miles with 5 bps/Hz spectral efficiency and its speed can goes up to 100 Mbps in a Flexible channel sizes from 1.5 MHz to 20 MHz. As such, WiMAX can bring the underlying Internet connection needed to service local Wi-Fi networks from one to hundreds of Consumer Premises Equipments (CPE)s, with unlimited subscribers behind each CPE (Tsolkas *et al.*, 2012). WiMax will provide your several levels of Quality of Service (QoS) and provides ubiquitous broadband.

Multi-carrier generation is the main building block for generating WiMax signal. In order to use WiMax signal in radio over fiber applications the use of all optical generation of RF signals can help to reduce the challenges of electronic devices. One solution to generate multi carriers optically is to use the MRR systems. Nonlinear light behaviour inside an MRR occurs when a strong pulse of light is inputted into the ring system, which is used in many applications in signal processing and communication (Spyropoulou *et al.*, 2011). The properties of a ring system can be modified via various control methods. Ring resonators can be used as filter devices where generation of high frequency (THz) soliton signals can be performed using suitable system parameters (Lin & Crozier, 2011). The series of MRRs connected to an add/drop filter system, is used in many applications in optical communication and signal processing. This system can be used to generate optical soliton pulses of THz frequencies, thus providing required signals used in wired/wireless optical communication such as all optical OFDM to be applied for WiMax applications.

In this section, series of ring resonators are connected to an add/drop system in order to generate multi-carriers which are applied to implement the optical OFDM suitable for the WiMax communication systems. Results show that MRR systems support both single and multi-carrier optical soliton pulses that are used in an OFDM transmitter/receiver system. Here, the optical soliton in a nonlinear fiber MRR system is analyzed in order to generate a high frequency band (THz) of pulses as single and multi-carrier signals where multi-carriers are used for generating one optical WiMax channel band. The generated signal is multiplexed with a single carrier soliton and transmitted through Single Mode Fiber (SMF) after being beaten to photodiode an IEEE802.16m signal is propagated wirelessly in transmitter antenna base station and is received by the second antenna located in the receiver. The bit error rate (BER) and error vector magnitude of the overall system are also discussed. The fixed and variable parameters of the MRR system are listed in Table 8.2.

The results of the chaotic signal generation are shown in Figure 8.18. The input pulse of the bright soliton pulse with a power of 800 mW is inserted into the system. Large bandwidth within the MRRs can be generated by using a bright soliton pulse input into the nonlinear system. The signal is chopped (sliced) into smaller signals spreading over the spectrum; thus, a large bandwidth is formed by the nonlinear effects of the medium. A frequency soliton pulse can be formed and trapped within the system with suitable ring parameters. The chaotic soliton pulses are used widely

Table 8.2 Fixed and variable parameters of the MRR system.

Fixed Parameters	Variable Parameters
$R_{Add/Drop} = 550\,\mu m$	$T_0 = $ Initial propagation time
$R_1 = 15\,\mu m$	$T = $ Propagation time
$R_2 = 9\,\mu m$	$z = $ Propagation distance
$R_3 = 6\,\mu m$	$L_D = $ Dispersion length
$R_4 = 6\,\mu m$	$L_{NL} = $ Nonlinear length
$\kappa_1 = 0.98$	$\phi = $ Total phase shift
$\kappa_2 = 0.98$	$\phi_{NL} = $ Nonlinear phase shift
$\kappa_3 = 0.96$	$\phi_0 = $ Linear phase shift
$\kappa_4 = 0.92$	$A = $ Optical amplitude
$\kappa_5 = 0.05$	$I = $ Optical intensity
$\kappa_6 = 0.05$	$P = $ Optical power
$n_0 = 3.34$	$E_1, E_2 = $ Interior electric fields
$n_2 = 2.4 \times 10^{-17}\,m^2\,W^{-1}$	$E_{out} = $ Electric field of the ring resonator
$A_{eff\,1} = 0.50\,\mu m^2$	$E_t = $ Throughput electric field
$A_{eff\,2} = 0.25\,\mu m^2$	$E_d = $ Drop port electric field
$A_{eff\,3} = 0.10\,\mu m^2$	
$\alpha = 0.5\,dBmm^{-1}$	
$\gamma = 0.1$	

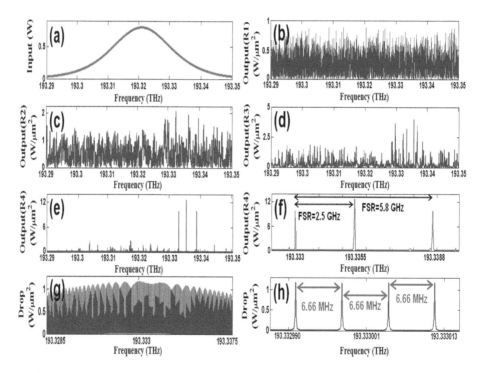

Figure 8.18 Results single and multi-carriers: (a) Input bright soliton, (b): Output from first ring, (c): Output from second ring, (d): Output from third ring, (e): Output from the fourth ring, (f): Expansion of the output R_4, (g): Drop port output, (h): Expansion of the output E_d.

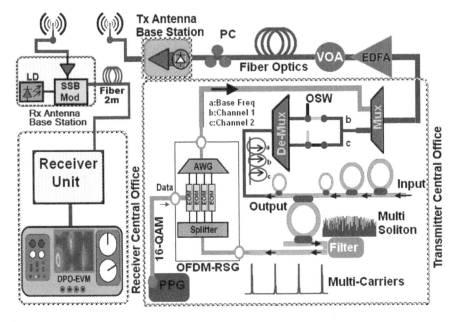

Figure 8.19 System setup.

as carrier signals in securing optical communication, wherein the information is input into the signals and ultimately can be retrieved by using suitable filtering systems.

Filtering of the soliton signals can be performed when the pulses pass through the MRRs. The output signals from the system can be seen in Figure 8.18, where soliton pulses ranges of 193.29–193.35 THz are generated and used in WiMax communication, Wireless Personal Area Networks (WPANs) and Wireless Local Area Networks (WLANs). The fourth MRR's output ($E_{out\,4}(t)$) shows localized ultra-short soliton pulses with FSR of 2.5 and 5.8 GHz, where soliton pulses at frequencies of 193.333, 193.3355 and 193.3388 GHz are generated. The drop port output expressed by E_d is shown in Figure 8.18 (g, h), where multi-soliton pulses with FSR of 6.66 MHz could be generated. For doing so the band pass filter is used to pick out the four carriers shown in Figure 8.18(h).

8.6.1   System setup

The schematic of the system setup is shown in Figure 8.19. At the Transmitter Central Office (TCO) a series of MRRs are connected to an add/drop system in order to generate single and multi-soliton carriers as discussed in the previous section.

8.6.2   Wimax signal generation

In this stage the WiMax signal is generated optically by using the OFDM signal generator module (OFDM-SGM). This module has one optical and one electrical input and one optical output port. The electrical input port is connected to the Pulse Pattern Generator (PPG) which generates 16-QAM data signal. Optical input is the multi-carriers centered shown in Figure (2h). In order to generate all optical OFDM signal, multi-carriers are first separated by the optical splitter then, they are modulated via 16-QAM signal from the PGG using External Optical Modulator (EOM). Afterwards, in order to characterize the IFFT block at the transmitter an array waveguide grating (AWG) is used (Wang *et al.*, 2011).

The spectra of the modulated optical subcarriers are overlapped which form one optical OFDM channel band. Then the generated all optical OFDM signal is multiplexed by the wavelength located at b:193.3355 or c:193.3388 i.e. channel 1 and channel 2. The distance between the center of OFDM signal and the center of the single carriers is (a to b) 2.5 GHz or (a to c) 5.8 GHz which are the RF band for WiMax standard.

These two channels are separated with de-multiplexer and using optical switch (OSW), one of them is switched into multiplexer with the base carrier (a) and after amplification by an Erbium Doped Fiber Amplifier (EDFA) the multiplexed signal is transmitted through the Single Mode Fiber (SMF). The nonlinear refractive index is 2.6×10^{-20} m^2/W, where the fiber optic has a length of $L = 10, 25,$ and 50 km, attenuation of 0.2 dB/km, dispersion of 5 ps/(nm.km), the differential group delay of 0.2 ps/km, effective area of 25 μm^2 and the nonlinear phase shift of 3 mrad.

In order to investigate the optical link performance, the total optical power level after amplification is adjusted with a Variable Optical Attenuator (VOA) from -3 to 7 dBm. The joint signal is received at the transmitter antenna base station and in order to maximize the photoreceiver performance, the state of the polarization can be adjusted by a Polarization Controller (PC). The multiplexed signals are being beaten to a PIN photodetector (0.7 A/W responsivity) therefore a RF WiMax signal is generated and propagated wirelessly by the transmitter antenna. Here, based on the switching of channel 1 or channel 2, whether 2.5 GHz or 5.8 GHz RF WiMax are generated and are shown in Figure 8.20(a) and 8.20(b) respectively.

At the receiver antenna base station, the propagated RF WiMax is received which is shown in Figure 8.20(c). Here the RF signal is up-converted using a commercially available Distributed Feedback (DFB) laser to process the received signal optically. Now the up-converted signal is transmitted to Receiver Central Office (RCO) through 2 m SMF. At the RCO the detected signal is amplified and analysed in order to evaluate the Error Vector Magnitude (EVM) of each wireless channel. EVM measurement as a figure of merit for assessing the quality of digital communication signals is performed to evaluate the link degradation.

The EVM results for channels 1 and 2 at different optical power and in different optical path lengths are shown in Figure 8.20(d). The -14.5 dB EVM is the threshold for successful transmission which is shown with a dashed line. As it is clear in results for different optical power both channels show a soft EVM variation for different path lengths. Therefore it could be concluded that the transmission of both channels is feasible for up to a 50 km SMF path length. A further investigation on system performance is conducted using a bit error rate calculation.

As illustrated in Figure 8.21, the system performance under two circumstances is investigated which are channel 1 and channel 2. As can be concluded from this figure by increasing the received power, channel 2 outperforms the other one.

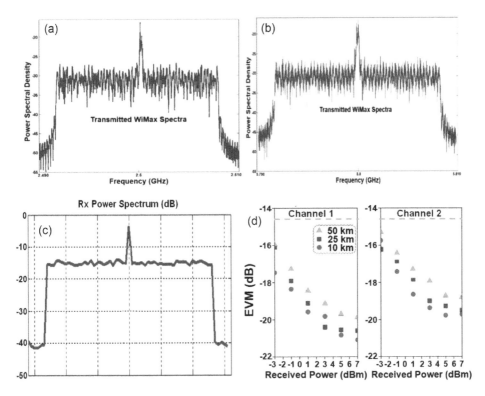

Figure 8.20   Transmitter and receiver performance.

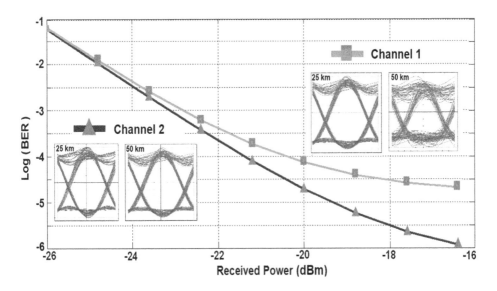

Figure 8.21   The system performance under two channel 1 and channel 2.

Based on the presented system and results it is possible to use MRR to generate both single and multi-carriers which can be applied to the optical generation of WiMax signal. IEEE802.16e is a common application which can be benefited from this system.

8.7 HIGHLY CHAOTIC SIGNAL GENERATION AND TRANSMISSION USING PANDA RING RESONATOR

The input pulses of the Gaussian pulse with power of 0.45 W are inserted into the PANDA ring resonator. The results of the chaotic signal are shown in Figure 8.22. The centered ring of the PANDA system has a radius of $100\,\mu$m, where the radii of the rings on the right and left sides are the same as $7\,\mu$m. The coupling coefficients of the PANDA system are selected to, $\kappa_1 = 0.7$, $\kappa_2 = 0.2$, $\kappa_0 = 0.01$ and $\kappa_3 = 0.85$. The nonlinear refractive index of the PANDA system is $n_2 = 1.3 \times 10^{-17}\,\text{m}^2\,\text{W}^{-1}$. Using the add port of the PANDA system, the signals can be amplified and tuned. The signals on the right side of the PANDA system are shown in Figure 8.22(a–b) where the Figure 8.22(c–d) shows the signals on the left side of the system.

More channel capacity can be obtained and controlled by generating large bandwidth of chaotic signals. Therefore, stable signals of the chaotic signals can be seen within the through port of the system shown in Figure 8.23.

The potential of chaotic bands can be generated and used for many applications such as optical trapping and coding-decoding telecommunication. Thus, the chaotic signals can be input into the transmission link to perform the optical trapping. The optical trapping transmission system of chaotic signals is shown in Figure 8.24.

In Figure 8.25, the fiber optic has a length of 195 km, attenuation of 0.4 dB/km, dispersion of 1.67 ps/(nm·km), the differential group delay of 0.2 ps/km, the nonlinear

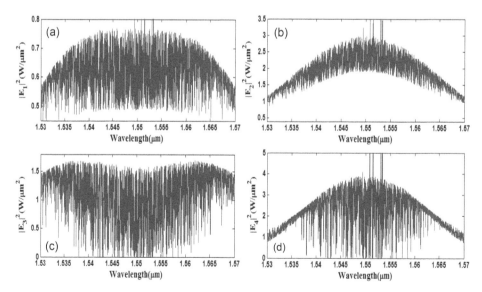

Figure 8.22 Chaotic signals generated by the PANDA system, where (a): $|E_1|^2$ (b): $|E_2|^2$, (c): $|E_3|^2$ and (d): $|E_4|^2$.

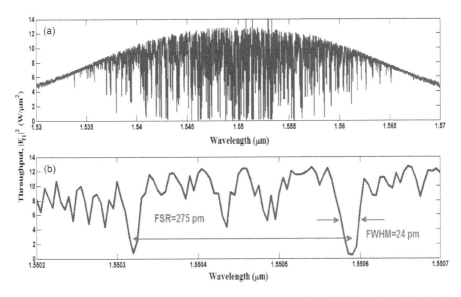

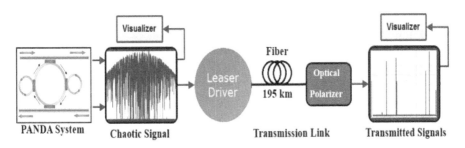

Figure 8.23 Chaotic signal generation using the PANDA system where (a): Throughput chaotic signals and (b): Expansion of the Throughput chaotic signals.

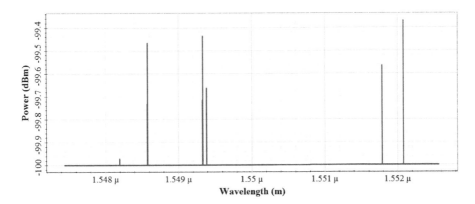

Figure 8.24 System of fiber optic transmission link.

Figure 8.25 Trapping of chaotic signals.

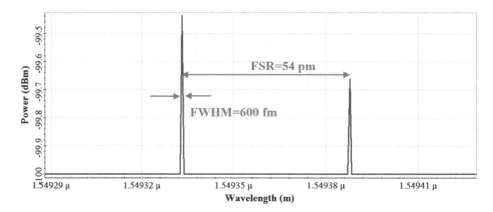

Figure 8.26 Transmitted chaotic signals with FWHM and FSR of 600 fm and 54 pm respectively.

refractive index of 2.6×10^{-20} m^2/W, effective area of 62.8 μm^2 and the nonlinear phase shift of 3 mrad. Figure 8.25 shows the trapping of chaotic signals in the communication system.

The trapping of signals can be obtained after the signals were transmitted along the fiber optic and finally received by suitable optical receiver thus detection process can be performed via the optical receiver. The Full Width at Half Maximum (FWHM) and Free Spectrum Range of the trapped signals can be seen in Figure 8.26. Here the pulses with FWHM and FSR of 600 fm and 45 pm could be generated experimentally.

Therefore, trapping of chaotic signals along the fiber optic is performed.

8.8   DARK SOLITON GENERATION AND TWEEZERS TRANSMISSION USING FIBER OPTIC LINK

The fixed and variable parameters of the Half-Panda system are listed in Table 8.3.

Input optical dark and bright solitons with powers 500 mW and 320 mW respectively are inserted into the Half-Panda system. The add-drop system has a radius of $R_{ad} = 15$ μm where the coupling coefficients are $\kappa_1 = \kappa_2 = 0.3$. The dark solitons are propagating inside the Half-Panda system with central wavelengths of $\lambda_0 = 1.4$ μm, 1.45 μm, 1.5 μm, 1.55 μm, 1.6 μm. In order to increase the capacity of the output optical tweezers, input dark soliton pulses with four different central wavelengths are introduced into the input port of the Half-Panda system. Figure 8.27(a), shows the optical inputs in the form of dark and bright soliton pulses. The nonlinear condition forms the interior signals as chaotic signals respect to 20,000 round-trip of the input power. In order to make the system associate with the practical device (InGaAsP/InP), the selected parameters of the system are fixed to $n_0 = 3.34$, $n_2 = 2.5 \times 10^{-17}$ and $A_{eff} = 25$ μm^2. By adjusting the parameters such as the dark and bright powers at

Table 8.3 Fixed and variable parameters of proposed MRR system.

Fixed Parameters	Variable Parameters
R_{ring} = ring's radius	T = Propagation time
R_{ad} = Add/drop ring's radius	z = Propagation distance
κ_1 = Add/drop coupling coefficients	L = Waveguide length
κ_2 = Add/drop coupling coefficients	L_D = Dispersion length
κ = ring's coupling coefficient	L_{NL} = Nonlinear length
L_{ring} = Circumference of the ring resonator	n = Total nonlinear refractive index
L_{ad} = Circumference of the add-drop ring	ϕ_{NL} = Nonlinear phase shift
n_0 = Linear refractive index	ϕ_0 = Linear phase shift
n_2 = Nonlinear refractive index	A = Optical amplitude
α = Ring resonator loss	E_0 = Electric field of the ring
β_1 = Propagation constant	I = Optical intensity
β_2 = Propagation constant	P = Optical power
γ = Coupler intensity loss	x = Round trip loss coefficient
A_{eff} = Effective core area	E_1 = Input electric field into the ring resonator
E_{i1} = Input electric field at input port	E_2 = Output electric field of the ring resonator
E_{i2} = Input electric field at drop port	E_3 = Left side's electric field of the of add-drop
ω_0 = Frequency carrier	$|E_{t1}|^2$ = Throughput output power
λ = Wavelength	$|E_{t2}|^2$ = Drop port output power
k = Wave propagation number	

the input and add ports and the coupling coefficients, the tweezers depth would be controlled and tuned as shown in Figure 8.27(b–d).

Smallest tweezers width of 8.85 nm is generated at the drop port shown in Figure 8.28, where the through port shows the output intensity signals with FWHM and FSR of 33 nm and 50 nm respectively. The cancelling of the chaotic signals can be obtained within the add-drop ring resonator using suitable parameters of the system. The signals can be controlled and tuned by power's variation of the input bright soliton pulse.

Transportation of the optical tweezers can be obtained via a network system using a transmission link. Detection of the transmitted tweezers signals can be assembled using the single photon detection method. Thus, the tweezers transportation for long distance communication via fiber optics is realistic. The system of optical tweezers transmission link is shown in Figure 8.29.

The fiber optic has a length of 100 km, attenuation of 0.2 dB/km, dispersion of 5 ps/(nm·km), the differential group delay of 0.2 ps/km, the nonlinear refractive index of 2.6×10^{-20} m²/W, effective area of 25 μm² and the nonlinear phase shift of 3 mrad. In operation, the signals can be modulated via an optical receiver unit which is encoded in the quantum signal transmission link. The receiver unit can be used to detect the transmitted optical tweezers. Transmitted optical tweezers can be sent to the users via a wired/wireless transmitter shown in Figure 8.30. The advanced transmitter topologies are desirable for application in both wired and wireless communication inasmuch as they are able to provide power-efficient amplification of signals with large Peak-to Average Power Ratios (PAPRs) without compromising system linearity. Figure 8.30 shows the detected and filtered optical tweezers using an optical receiver.

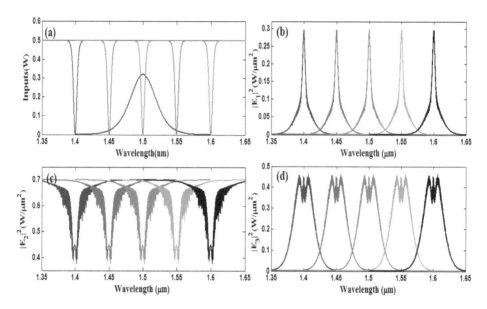

Figure 8.27   Optical tweezers generation within a Half-Panda system where (a): input optical dark and bright solitons, (b–d): tuned optical tweezers.

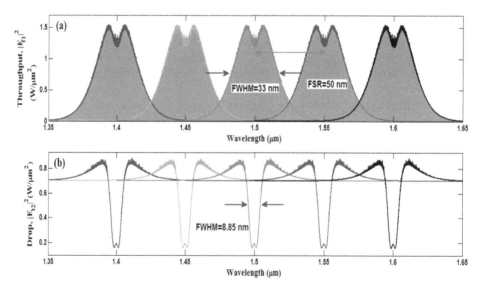

Figure 8.28   Through and drop port output signals of the Half-Panda system where (a): Through port output with FWHM = 33 nm, (b): drop port output with FWHM = 8.85 nm.

By using suitable dark-bright soliton input powers, tunable optical tweezers can be controlled. High capacity data transmission can be applied by using more wavelengths. The advantage of this study is that optical tweezers can be generated and transmitted via a network system thus improving the transmission capacity.

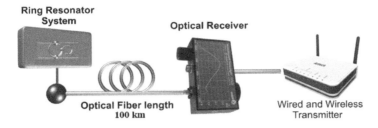

Figure 8.29 System of optical tweezers transmission link.

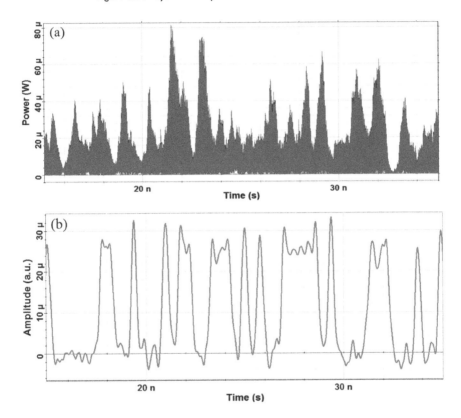

Figure 8.30 Detected and filtered optical tweezers via a 100 km optical fiber optic using an optical receiver, where (a): detected signals, (b): filtered signals.

8.9 QUANTUM ENTANGLED PHOTONS GENERATION BY TWEEZERS AND TRANSMISSION USING THE WIRELESS ACCESS POINT SYSTEM

Input optical dark solitons and Gaussian laser bean with powers 2 W and 1 W respectively are inserted into the Half-Panda system. The add-drop optical filter has radius of $R_{ad} = 15\,\mu$m where the coupling coefficients are $\kappa_1 = 0.35$ and $\kappa_2 = 0.25$. The dark

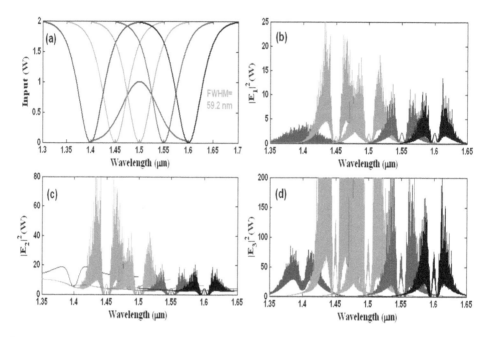

Figure 8.31   Optical tweezers generation within a Half-Panda system where (a): input of optical dark solitons and Gaussian laser beam, (b–d): amplified and tuned optical tweezers in the form of potential wells.

solitons are propagating inside the Half-Panda system with central wavelengths of $\lambda_0 = 1.4\,\mu m$, $1.45\,\mu m$, $1.5\,\mu m$, $1.55\,\mu m$, $1.6\,\mu m$. In order to increase the capacity of the output optical tweezers, input dark soliton pulses with four different central wavelengths are introduced to the input port of the system. Figure 8.31(a), shows the optical inputs in the form of dark soliton and Gaussian laser beam.

The nonlinear condition forms the interior signals as chaotic signals respect to 20,000 roundtrips of the input power. In order to make the system associate with the practical device (InGaAsP/InP), the selected parameters of the system are fixed to $n_0 = 3.34$ and $n_2 = 2.5 \times 10^{-17}$. By adjusting the parameters such as the dark and Gaussian powers at the input and add ports and the coupling coefficients, the tweezers depth would be controlled and tuned as shown in Figure 8.31(b–d). Amplification of the signals occurs within the nonlinear system which makes the signals suitable for long distance communication.

Smallest tweezers width of 4.2 nm is generated at the through port shown in Figure 8.32, where the drop port shows the output signal with FWHM of 18.5 nm. Transportation of the optical tweezers can be obtained via a network system using a photon generator and transmitter. Detection of the transported tweezers signal can be assembled using the single photon detection method. Thus, the tweezers transportation for long distance communication via molecular transporter is realistic.

Cancelling of the chaotic signals can be obtained within the add-drop ring resonator interferometer system using suitable parameters of the rings. The signals can

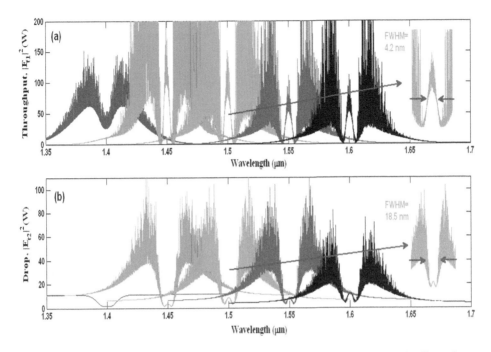

Figure 8.32 Through and drop port output signals of the Half-Panda system where (a): Through port output with FWHM = 4.2 nm, (b): drop port output with FWHM = 18.5 nm.

be controlled and tuned by power's variation of the input Gaussian laser pulse. Figure 8.33 shows the generation of nano optical tweezers (clear signals). Here the input powers of the optical dark soliton pulses and Gaussian laser beam are 2 W and 2.5 W respectively.

Filtered and clear optical tweezers are seen in Figure 8.34 where the peaks have FWHM and FSR of 8.9 nm and 50 nm respectively. In the case of communication networks, generation of narrower signals is recommended. Therefore soliton signals can be used in optical communication where the capacity of the output signals can be improved by generation of peaks with smaller FWHM. The sensitivity of the ring systems can be improved significantly by generation of peaks with wider space or bigger FSR.

The proposed transmission unit is a quantum processing system that can be used to generate high capacity packet of quantum entanglement photons within the series of MRRs. In operation, the computing data can be modulated and input into the system via a receiver unit which is encoded to the quantum signal processing system. The receiver unit can be used to detect the quantum bits. It is obtained via the reference states recognized by using the cloning unit operated by an add-drop filter (R_{d1}) shown in Figure 8.35.

By using suitable dark-Gaussian soliton input power, tunable optical tweezers can be controlled (Ali *et al.*, 2010a, 2010k). This provides the entangled photon as the dynamic optical tweezers probe. The required data can be retrieved via the through

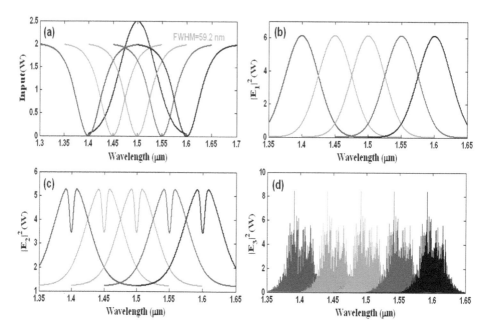

Figure 8.33 Optical tweezers generation within a Half-Panda system where (a): input of dark solitons and Gaussian laser beam, (b–d): tuned and controlled optical tweezers.

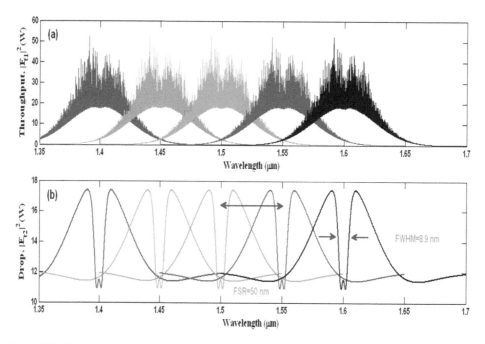

Figure 8.34 Through and drop port output signals of the Half-Panda system where (a): Through port chaotic output signals (b): drop port output with FWHM = 8.9 nm and FSR = 50 nm.

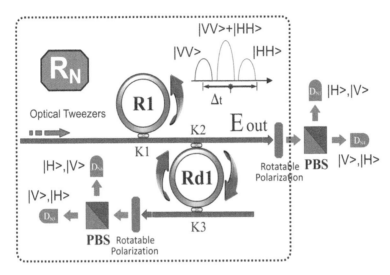

Figure 8.35 Schematic of entangled photon manipulation system. The quantum state is propagating to a rotatable polarizer and then is split by a beam splitter (PBS) flying to detector D_{N3}, D_{N4}, D_{N5} and D_{N6}.

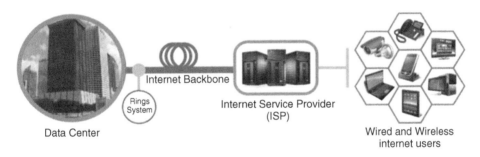

Figure 8.36 System of entangled photons transmission using a wireless access point system.

and drop ports of the add-drop filter in the router. High capacity data transmission can be applied by using more wavelength carriers. The advantage of this study is that ultra-short nano optical tweezers can be generated and transmitted via a network system thus improving the security and capacity.

The polarization states of light pulses are changed and converted during the circulation in the delay circuit, leading to the formation of the polarized entangled photon pairs. The entangled photons of the nonlinear ring resonator are then separated into the signal and idler photon probability. The polarization angle adjustment device is applied to investigate the orientation and optical output intensity.

Transporter states can be controlled and identified using the quantum processing system as shown in Figure 8.35. In operation, the encoded quantum secret codes computing data can be modulated and input into the system via a wireless router. Schematic of the wireless router is shown in Figure 8.36, in which quantum cryptography for

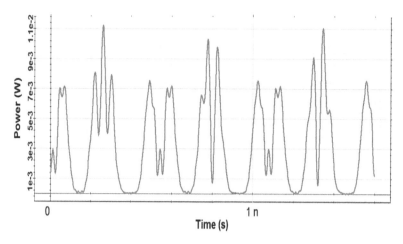

Figure 8.37 Detected and filtered optical tweezers via a 25 m optical wireless link using an optical receiver.

internet security can be obtained. A wireless router system can be used to transfer generated entangled photons via a wireless access point, and network communication system shown in Figure 8.36.

A wireless access system transmits data to different users via wireless connection. This method also works in reverse, when the router system used to receive information from the Internet, translating it into an analog signal and sending it to the computer's wireless adapter. Figure 8.37 shows the detected and filtered optical tweezers using an optical receiver.

The advanced transmitter topologies are desirable for application in both wired and wireless communication inasmuch as they are able to provide power-efficient amplification of signals with large Peak-to Average Power Ratios (PAPRs) without compromising system linearity.

Chapter 9

Conclusion

Chaotic signals can be generated using the input laser power propagating within a nonlinear ring resonator, where the required signals of single bandwidth soliton pulse can be recovered and manipulated by using an add/drop system. Results obtained have shown particular possibilities with this application. Also, effects of coupling coefficients on the bandwidth of the single soliton pulse have been presented here. The add/drop system is the efficient system to generate the soliton comb which has several applications in optical communication.

The PANDA is presented as optical chaos. The Gaussian beams with center wavelength of $1.55\,\mu m$, are inserted into the system which are good to generate a high capacity of chaotic signals. Transmission of chaotic signals can be obtained via a fiber optic communication link with the length of 195 km, where trapping of the signals can be performed. Here the trapped signals with FHWM $= 600$ fm is generated. Nano bandwidth optical tweezers can be generated by the dark soliton propagation in a Half-Panda system. The control process of the dark-bright collision can be performed by using suitable parameters of the ring system such as the input power, coupling coefficient, ring radius, coupler loss and effective core area.

Tweezers with FWHM and FSR of 33 nm and 50 nm respectively can be transmitted experimentally via a fiber optic with a length of 100 km. The results show the detected tweezers, where the clear signals can be obtained by cancelling the chaotic signals. Tweezers can be used in cryptography and secured optical communication. Nano optical tweezers can be generated by the dark soliton propagation in a Half-Panda system. The control process of the dark-Gaussian collision can be implemented by using suitable parameters of the ring system The Quantum signal processing unit is connected to the optical tweezers which can be used to generate quantum entanglement photons thus providing secured and high capacity information.

In the case of MRRs, the measured normalized transmitted light at the through port for both TE (left) and TM (right) modes using a tunable source in the $\lambda = 1.55$ to $1.57\,\mu m$ region have been presented. The resonance of a TM mode shows a $\Delta\lambda = 105$ pm-shift in response to an applied voltage of $\Delta V = 100$ V. This wavelength shift corresponds to frequency tunability of 0.14 GHz/V.

This secured coded information can be easily transmitted via a communication network system. Here Double security can be performed when ultra-short of soliton pulses are used to generate entangle photon transmitting over long distance communication. The slow light generation is studied by simulation in linear and nonlinear micro-ring resonator configurations for three different optical waveguide materials such as InGaAsP/InP, GaAlAs/GaAs and hydrogenated amorphous silicon. The bright soliton and Gaussian beam is used as input pulse. The mathematical derivations for the proposed micro-ring resonator configurations for the generation of slow light are performed using scattering matrix method and simulated by using MATLAB programming. The slow light generation is realized by proposed micro-ring resonator system. The key findings of present research work are given as micro-ring resonator configurations for three different optical waveguide materials such as InGaAsP/InP, GaAlAs/GaAs and hydrogenated amorphous silicon. The bright soliton and Gaussian beam is used as input pulse. The mathematical derivations for the proposed micro-ring resonator configurations for the generation of slow light are performed using scattering matrix method and simulated by using MATLAB programming. The slow light generation is realized by proposed micro ring resonator system. Therefore, the slow light pulse is generated for three different waveguides for linear and nonlinear microring configurations. In the nonlinear configurations of micro-ring resonators, the time delay in output signals are obtained 46.42 ns, 45.36 ns and 48.71 ns for InGaAsP/InP, GaAlAs/GaAs and hydrogenated amorphous silicon waveguides respectively for bright soliton as an input pulse.

References

Abdullaev, F. and Garnier, J. (2005) Optical solitons in random media. *Progress in Optics*, 48: 35–106.

Absil, P., Hryniewicz, J., Little, B., Cho, P., Wilson, R., Joneckis, L. and Ho, P.T. (2000) Wavelength conversion in GaAs micro-ring resonators. *Optics Letters*, 25(8): 554–556.

Afroozeh, A., Amiri, I.S. and Zeinalinezhad, A. (2014) *Micro Ring Resonators and Applications*. Saarbrücken, Germany, LAP LAMBERT Academic Publishing.

Afroozeh, A., Amiri, I.S., Ali, J. and Yupapin, P.P. (2012a) Determination of Fwhm for Solition Trapping. *Jurnal Teknologi (Sciences and Engineering)*, 55: 77–83.

Afroozeh, A., Amiri, I.S., Bahadoran, M., Ali, J. and Yupapin, P.P. (2012b) Simulation of Soliton Amplification in Micro Ring Resonator for Optical Communication. *Jurnal Teknologi (Sciences and Engineering)*, 55: 271–277.

Afroozeh, A., Amiri, I.S., Chaudhary, K., Ali, J. and Yupapin, P.P. (2014b) Analysis of Optical Ring Resonator. *Advances in Laser and Optics Research*. New York, Nova Science.

Afroozeh, A., Amiri, I.S., Jalil, M.A., Kouhnavard, M., Ali, J. and Yupapin, P.P. (2011a) Multi Soliton Generation for Enhance Optical Communication. *Applied Mechanics and Materials*, 83: 136–140.

Afroozeh, A., Amiri, I.S., Kouhnavard, M., Bahadoran, M., Jalil, M.A., Ali, J. and Yupapin, P.P. (2010a) Dark and Bright Soliton trapping using NMRR. *International Conference on Experimental Mechanics (ICEM)*. Kuala Lumpur, Malaysia.

Afroozeh, A., Amiri, I.S., Kouhnavard, M., Bahadoran, M., Jalil, M.A., Ali, J. and Yupapin, P.P. (2010b) Optical Memory Time using Multi Bright Soliton. *International Conference on Experimental Mechanics (ICEM)*. Kuala Lumpur, Malaysia.

Afroozeh, A., Amiri, I.S., Kouhnavard, M., Jalil, M., Ali, J. and Yupapin, P. (2010c) Optical dark and bright soliton generation and amplification. *AIP Conference Proceedings*, 1341: 259–263.

Afroozeh, A., Amiri, I.S., Zeinalinezhad, A., Pourmand, S.E. and Ahmad, H. (2015) Comparison of Control Light using Kramers-Kronig Method by Three Waveguides. *Journal of Computational and Theoretical Nanoscience (CTN)*.

Afroozeh, A., Bahadoran, M., Amiri, I.S., Samavati, A.R., Ali, J. and Yupapin, P.P. (2011b) Fast Light Generation Using Microring Resonators for Optical Communication. *National Science Postgraduate Conference, NSPC*. Universiti Teknologi Malaysia.

Afroozeh, A., Pourmand, S.E. and Zeinalinezhad, A. (2014d) Simulation and Calculation of the Parameters of PCF under Terahertz Wave Band. *Advances in Environmental Biology*, 8(21): 5.

Afroozeh, A., Zeinalinezhad. A. and Pourmand, S.E. (2014a) Evaluation of The Protein of Nanostructured Silica Based Biomaterial. *IJBPAS*, 3(11): 6.

Afroozeh, A., Zeinalinezhad, A., Amiri, I.S. and Pourmand, S.E. (2014a) Stop Light Generation using Nano Ring Resonators for ROM. *Journal of Computational and Theoretical Nanoscience (CTN)*, 12(3).

Afroozeh, A., Zeinalinezhad. A., Pourmand, S.E. and Amiri, I.S. (2014) Determination of Suitable Material to Control of Light. *IJBPAS*, 3(11): 11.

Afroozeh, A., Zeinalinezhad. A., Pourmand, S.E. and Amiri, I.S. (2014b) Attosecond pulse generation using nano ring waveguides. *IJCLS*, 5(9): 5.

Afroozeh, A., Zeinalinezhad. A., Pourmand, S.E. and Amiri, I.S. (2014c) Optical Soliton Pulse Generation to Removal of Tissue. *IJBPAS*, 3(11).

Agrawal, G. (2000) Nonlinear fiber optics. *Nonlinear Science at the Dawn of the 21st Century*, 195–211, Springer.

Ahn, J., Fiorentino, M., Beausoleil, R.G., Binkert, N., Davis, A., Fattal, D., Jouppi, N.P., McLaren, M., Santori, C.M. and Schreiber, R.S. (2009) Devices and architectures for photonic chip-scale integration. *Applied Physics A*, 95(4): 989–997.

Aitchison, J., Al-Hemyari, K., Ironside, C., Grant, R. and Sibbett, W. (1992) Observation of spatial solitons in AlGaAs waveguides. *Electronics Letters*, 28(20): 1879–1880.

Akanbi, O.A., Amiri, I.S. and Fazeldehkordi, E. (2015) *A Machine Learning Approach to Phishing Detection and Defense*. USA, Elsevier.

Alavi, S., Amiri, I., Idrus, S., Supa'at, A., Ali, J. and Yupapin, P. (2014) All Optical OFDM Generation for IEEE802.11a Based on Soliton Carriers Using MicroRing Resonators. *IEEE Photonics Journal*, 6.

Alavi, S.E., Amiri, I.S. and Supa'at, A.S.M. (2014) *Analysis of VFSO System Integrated With BPLC*. Amazon, Lap Lambert Academic Publishing.

Alavi, S.E., Amiri, I.S., Ahmad, H., Fisal, N. and Supa'at, A.S.M. (2015a) Optical Amplification of Tweezers and Bright Soliton Using an Interferometer Ring Resonator System. *Journal of Computational and Theoretical Nanoscience (CTN)*, 12(4).

Alavi, S.E., Amiri, I.S., Ahmad, H., Supa'at, A.S.M. and Fisal, N. (2014) Generation and Transmission of 3 × 3 W-Band MIMO-OFDM-RoF Signals Using Micro-Ring Resonators. *Applied Optics*, 53(34): 8049–8054.

Alavi, S.E., Amiri, I.S., Idrus, S.M. and Ali, J. (2013b) Optical Wired/Wireless Communication Using Soliton Optical Tweezers. *Life Science Journal*, 10(12s): 179–187.

Alavi, S.E., Amiri, I.S., Idrus, S.M. and Supa'at, A.S.M. (2014a) Generation and Wired/Wireless Transmission of IEEE802.16m Signal Using Solitons Generated By Microring Resonator. *Optical and Quantum Electronics*.

Alavi, S.E., Amiri, I.S., Idrus, S.M., Supa'at, A.S.M. and Ali, J. (2015c) Cold Laser Therapy Modeling of Human Cell/Tissue by Soliton Tweezers. *Optik*.

Alavi, S.E., Amiri, I.S., Idrus, S.M., Supa'at, A.S.M. and Ali, J. (2013a) Chaotic Signal Generation and Trapping Using an Optical Transmission Link. *Life Science Journal*, 10(9s): 186–192.

Alavi, S.E., Amiri, I.S., Idrus, S.M., Supa'at, A.S.M., Ali, J. and Yupapin, P.P. (2014b) All Optical OFDM Generation for IEEE802.11a Based on Soliton Carriers Using MicroRing Resonators *IEEE Photonics Journal*, 6(1).

Alavi, S.E., Amiri, I.S., Soltanian, M.R.K., Supa'at, A.S.M., Fisal, N. and Ahmad, H. (2015b) Generation of Femtosecond Soliton Tweezers Using a Half-Panda System for Modeling the Trapping of a Human Red Blood Cell. *Journal of Computational and Theoretical Nanoscience (CTN)*, 12(1).

Alavi, S.E., Amiri, I.S., Supa'at, A.S.M. and Idrus, S.M. (2015d) Indoor Data Transmission Over Ubiquitous Infrastructure of Powerline Cables and LED Lighting. *Journal of Computational and Theoretical Nanoscience (CTN)*.

Ali, J., Afroozeh, A., Amiri, I.S., Hamdi, M., Jalil, M., Kouhnavard, M. and Yupapin, P. (2010a) Entangled photon generation and recovery via MRR. *ICAMN, International Conference.* Prince Hotel, Kuala Lumpur, Malaysia.

Ali, J., Afroozeh, A., Amiri, I.S., Jalil, M. and Yupapin, P. (2010b) Wide and narrow signal generation using chaotic wave. *Nanotech Malaysia, International Conference on Enabling Science & Technology.* Kuala Lumpur, Malaysia.

Ali, J., Afroozeh, A., Amiri, I.S., Jalil, M.A. and Yupapin, P.P. (2010c) Dark and Bright Soliton trapping using NMRR.*ICEM.* Legend Hotel, Kuala Lumpur, Malaysia.

Ali, J., Afroozeh, A., Hamdi, M., Amiri, I.S., Jalil, M.A., Kouhnavard, M. and Yupapin, P. (2010d) Optical bistability behaviour in a double-coupler ring resonator. *ICAMN, International Conference.* Prince Hotel, Kuala Lumpur, Malaysia.

Ali, J., Amiri, I.S., Afroozeh, A., Kouhnavard, M., Jalil, M. and Yupapin, P. (2010e) Simultaneous dark and bright soliton trapping using nonlinear MRR and NRR. *ICAMN, International Conference.* Prince Hotel, Kuala Lumpur, Malaysia.

Ali, J., Amiri, I.S., Jalil, A., Kouhnavard, A., Mitatha, B. and Yupapin, P. (2010f) Quantum internet via a quantum processor. *International Conference on Photonics (ICP 2010).* Langkawi, Malaysia.

Ali, J., Amiri, I.S., Jalil, M., Kouhnavard, M., Afroozeh, A., Naim, I. and Yupapin, P. (2010g) Narrow UV pulse generation using MRR and NRR system. *ICAMN, International Conference.* Prince Hotel, Kuala Lumpur, Malaysia.

Ali, J., Amiri, I.S., Jalil, M.A., Afroozeh, A., Kouhnavard, M. and Yupapin, P. (2010h) Novel system of fast and slow light generation using micro and nano ring resonators. *ICAMN, International Conference.* Prince Hotel, Kuala Lumpur, Malaysia.

Ali, J., Amiri, I.S., Jalil, M.A., Hamdi, M., Mohamad, F.K., Ridha, N.J. and Yupapin, P.P. (2010i) Proposed molecule transporter system for qubits generation. *Nanotech Malaysia, International Conference on Enabling Science & Technology.* Malaysia.

Ali, J., Amiri, I.S., Jalil, M.A., Hamdi, M., Mohamad, F.K., Ridha, N.J. and Yupapin, P.P. (2010j) Trapping spatial and temporal soliton system for entangled photon encoding. *Nanotech Malaysia, International Conference on Enabling Science & Technology.* Kuala Lumpur, Malaysia.

Ali, J., Aziz, M., Amiri, I.S., Jalil, M., Afroozeh, A., Nawi, I. and Yupapin, P. (2010k) Soliton wavelength division in MRR and NRR Systems. *AMN-APLOC International Conference.* Wuhan, China.

Ali, J., Jalil, M., Amiri, I.S., Afroozeh, A., Kouhnavard, M., Naim, I. and Yupapin, P. (2010l) Multi-wavelength narrow pulse generation using MRR. *ICAMN, International Conference.* Prince Hotel, Kuala Lumpur, Malaysia.

Ali, J., Jalil, M.A., Amiri, I.S., Afroozeh, A., Kouhnavard, M. and Yupapin, P.P. (2010m) Generation of tunable dynamic tweezers using dark-bright collision. *ICAMN, International Conference.* Prince Hotel, Kuala Lumpur, Malaysia.

Ali, J., Jalil, M.A., Amiri, I.S. and Yupapin, P.P. (2010n) Dark-bright solitons conversion system via an add/drop filter for signal security application.*ICEM.* Legend Hotel, Kuala Lumpur, Malaysia.

Ali, J., Jalil, M.A., Amiri, I.S. and Yupapin, P.P. (2010o) MRR quantum dense coding. *Nanotech Malaysia, International Conference on Enabling Science & Technology.* KLCC, Kuala Lumpur, Malaysia.

Ali, J., Kouhnavard, M., Amiri, I.S., Afroozeh, A., Jalil, M.A., Naim, I. and Yupapin, P.P. (2010p) Localization of soliton pulse using nano-waveguide. *ICAMN, International Conference.* Prince Hotel, Kuala Lumpur, Malaysia.

Ali, J., Kouhnavard, M., Amiri, I.S., Jalil, M.A., Afroozeh, A. and Yupapin, P.P. (2010q) Security confirmation using temporal dark and bright soliton via nonlinear system. *ICAMN, International Conference*. Prince Hotel, Kuala Lumpur, Malaysia.

Ali, J., Kouhnavard, M., Jalil, M.A. and Amiri, I.S. (2010r) Quantum signal processing via an optical potential well. *Nanotech Malaysia, International Conference on Enabling Science & Technology*. Kuala Lumpur, Malaysia.

Ali, J., Kulsirirat, K., Techithdeera, W., Jalil, M.A., Amiri, I.S., Naim, I. and Yupapin, P.P. (2010s) Temporal dark soliton behavior within multi-ring resonators. *Nanotech Malaysia, International Conference on Enabling Science & Technology*. Malaysia.

Ali, J., Mohamad, A., Nawi, I., Amiri, I.S., Jalil, M., Afroozeh, A. and Yupapin, P. (2010t) Stopping a dark soliton pulse within an NNRR. *AMN-APLOC International Conference*. Wuhan, China.

Ali, J., Nur, H., Lee, S., Afroozeh, A., Amiri, I.S., Jalil, M., Mohamad, A. and Yupapin, P. (2010u) Short and millimeter optical soliton generation using dark and bright soliton. *AMN-APLOC International Conference*. Wuhan, China.

Ali, J., Raman, K., Afroozeh, A., Amiri, I.S., Jalil, M.A., Nawi, I.N. and Yupapin, P.P. (2010v) Generation of DSA for security application. *2nd International Science, Social Science, Engineering Energy Conference (I-SEEC 2010)*. Nakhonphanom, Thailand.

Ali, J., Raman, K., Kouhnavard, M., Amiri, I.S., Jalil, M.A., Afroozeh, A. and Yupapin, P.P. (2011) Dark soliton array for communication security. *AMN-APLOC International Conference*. Wuhan, China.

Al-Raweshidy, H. and Komaki, S. (2002) *Radio over fiber technologies for mobile communications networks*. Artech House Publishers.

Amiri, I.S. (2011a) FWHM Measurement of Localized Optical Soliton. *The International Conference for Nano materials Synthesis and Characterization* Malaysia, International Atomic Energy Agency (IAEA).

Amiri, I.S. (2011b) Optical Soliton Trapping for Quantum Key Generation. *The International Conference for Nano materials Synthesis and Characterization* Malaysia, International Atomic Energy Agency (IAEA).

Amiri, I.S. (2014) *Light Detection and Ranging Using NIR (810 nm) Laser Source*. Germany, LAP LAMBERT Academic Publishing.

Amiri, I.S. and Afroozeh, A. (2014) *Ring Resonator Systems to Perform the Optical Communication Enhancement Using Soliton*. USA, Springer.

Amiri, I.S. and Afroozeh, A. (2014) Spatial and Temporal Soliton Pulse Generation By Transmission of Chaotic Signals Using Fiber Optic Link *Advances in Laser and Optics Research*. New York, Nova Science Publisher. 11.

Amiri, I.S. and Afroozeh, A. (2014a) Integrated Ring Resonator Systems.*Ring Resonator Systems to Perform Optical Communication Enhancement Using Soliton*. USA, Springer.

Amiri, I.S. and Afroozeh, A. (2014b) Introduction of Soliton Generation.*Ring Resonator Systems to Perform Optical Communication Enhancement Using Soliton*. USA, Springer.

Amiri, I.S. and Afroozeh, A. (2014c) Mathematics of Soliton Transmission in Optical Fiber.*Ring Resonator Systems to Perform Optical Communication Enhancement Using Soliton*. USA, Springer.

Amiri, I.S. and Afroozeh, A. (2014d) Soliton Generation Based Optical Communication.*Ring Resonator Systems to Perform Optical Communication Enhancement Using Soliton*. USA, Springer.

Amiri, I.S. and Ahmad, H. (2014) *Optical Soliton Communication Using Ultra-Short Pulses*. USA, Springer.

Amiri, I.S. and Ali, J. (2012) Generation of Nano Optical Tweezers Using an Add/drop Interferometer System. *2nd Postgraduate Student Conference (PGSC)*. Singapore.

Amiri, I.S. and Ali, J. (2013) Nano Particle Trapping By Ultra-short tweezer and wells Using MRR Interferometer System for Spectroscopy Application. *Nanoscience and Nanotechnology Letters*, 5(8): 850–856.

Amiri, I.S. and Ali, J. (2013a) Controlling Nonlinear Behavior of a SMRR for Network System Engineering. *International Journal of Engineering Research and Technology (IJERT)*, 2(2).

Amiri, I.S. and Ali, J. (2013b) Data Signal Processing via a Manchester Coding-Decoding Method Using Chaotic Signals Generated by a PANDA Ring Resonator. *Chinese Optics Letters*, 11(4): 041901(041904).

Amiri, I.S. and Ali, J. (2013c) Nano Optical Tweezers Generation Used for Heat Surgery of a Human Tissue Cancer Cells Using Add/Drop Interferometer System. *Quantum Matter*, 2(6): 489–493.

Amiri, I.S. and Ali, J. (2013d) Optical Buffer Application Used for Tissue Surgery Using Direct Interaction of Nano Optical Tweezers with Nano Cells. *Quantum Matter*, 2(6): 484–488.

Amiri, I.S. and Ali, J. (2013e) Single and Multi Optical Soliton Light Trapping and Switching Using Microring Resonator. *Quantum Matter*, 2(2): 116–121.

Amiri, I.S. and Ali, J. (2014a) Characterization of Optical Bistability in a Fiber Optic Ring Resonator. *Quantum Matter*, 3(1): 47–51.

Amiri, I.S. and Ali, J. (2014a) Femtosecond Optical Quantum Memory generation Using Optical Bright Soliton. *Journal of Computational and Theoretical Nanoscience (CTN)*, 11(6): 1480–1485.

Amiri, I.S. and Ali, J. (2014b) Deform of Biological Human Tissue Using Inserted Force Applied by Optical Tweezers Generated by PANDA Ring Resonator. *Quantum Matter*, 3(1): 24–28.

Amiri, I.S. and Ali, J. (2014b) Generating Highly Dark–Bright Solitons by Gaussian Beam Propagation in a PANDA Ring Resonator. *Journal of Computational and Theoretical Nanoscience (CTN)*, 11(4): 1092–1099.

Amiri, I.S. and Ali, J. (2014c) Optical Quantum Generation and Transmission of 57–61 GHz Frequency Band Using an Optical Fiber Optics *Journal of Computational and Theoretical Nanoscience (CTN)*, 11(10): 2130–2135.

Amiri, I.S. and Ali, J. (2014c) Picosecond Soliton pulse Generation Using a PANDA System for Solar Cells Fabrication. *Journal of Computational and Theoretical Nanoscience (CTN)*, 11(3): 693–701.

Amiri, I.S. and Ali, J. (2014d) Simulation of the Single Ring Resonator Based on the Z-transform Method Theory. *Quantum Matter*, 3(6): 519–522.

Amiri, I.S. and Naraei, P. (2014) Optical Transmission Characteristics of an Optical Add-Drop Interferometer System. *Quantum Matter*.

Amiri, I.S. and Nikoukar, A. (2010–2011) Secured Binary Codes Generation for Computer Network Communication. *Network Technologies & Communications (NTC) Conference*. Singapore.

Amiri, I.S. and Shahidinejad, A. (2014) Generating of 57–61 GHz Frequency Band Using a Panda Ring Resonator *Quantum Matter*.

Amiri, I.S., Afroozeh, A. and Bahadoran, M. (2011a) Simulation and Analysis of Multisoliton Generation Using a PANDA Ring Resonator System. *Chinese Physics Letters*, 28(10): 104205.

Amiri, I.S., Afroozeh, A. and Pourmand, S.E. (2015) *Control of Light Using Microring Resonators*. USA, Elsevier.

Amiri, I.S., Afroozeh, A., Bahadoran, M., Ali, J. and Yupapin, P.P. (2011b) Up and Down Link of Soliton for Network Communication. *National Science Postgraduate Conference, NSPC.* Universiti Teknologi Malaysia.

Amiri, I.S., Afroozeh, A., Bahadoran, M., Ali, J. and Yupapin, P.P. (2012a) Molecular Transporter System for Qubits Generation. *Jurnal Teknologi (Sciences and Engineering)*, 55: 155–165.

Amiri, I.S., Afroozeh, A., Nawi, I.N., Jalil, M.A., Mohamad, A., Ali, J. and Yupapin, P.P. (2011c) Dark Soliton Array for communication security. *Procedia Engineering*, 8: 417–422.

Amiri, I.S., Ahmad, H. and Zulkifli, M.Z. (2014) Integrated ring resonator system analysis to Optimize the soliton transmission. *International Research Journal of Nanoscience and Nanotechnology*, 1(1): 002–007.

Amiri, I.S., Ahsan, R., Shahidinejad, A., Ali, J. and Yupapin, P.P. (2012b) Characterisation of bifurcation and chaos in silicon microring resonator. *IET Communications*, 6(16): 2671–2675.

Amiri, I.S., Alavi, S.E. and Ahmad, H. (2015a) Analytical Treatment of the Ring Resonator Passive Systems and Bandwidth Characterization Using Directional Coupling Coefficients *Journal of Computational and Theoretical Nanoscience (CTN)*, 12(3).

Amiri, I.S., Alavi, S.E. and Ahmad, H. (2015b) RF signal generation and wireless transmission using PANDA and Add/drop systems. *Journal of Computational and Theoretical Nanoscience (CTN)*.

Amiri, I.S., Alavi, S.E. and Ali, J. (2013b) High Capacity Soliton Transmission for Indoor and Outdoor Communications Using Integrated Ring Resonators. *International Journal of Communication Systems*, 28(1): 147–160.

Amiri, I.S., Alavi, S.E. and Ali, J. (2013c) High Capacity Soliton Transmission for Indoor and Outdoor Communications Using Integrated Ring Resonators. *International Journal of Communication Systems*.

Amiri, I.S., Alavi, S.E. and Idrus, S.M. (2014d) *Soliton Coding for Secured Optical Communication Link*. USA, Springer.

Amiri, I.S., Alavi, S.E. and Idrus, S.M. (2015a) Introduction of Fiber Waveguide and Soliton Signals Used to Enhance the Communication Security. *Soliton Coding for Secured Optical Communication Link*, 1–16. USA, Springer.

Amiri, I.S., Alavi, S.E. and Idrus, S.M. (2015b) Results of Digital Soliton Pulse Generation and Transmission Using Microring Resonators. *Soliton Coding for Secured Optical Communication Link*, 41–56. USA, Springer.

Amiri, I.S., Alavi, S.E. and Idrus, S.M. (2015c) Theoretical Background of Microring Resonator Systems and Soliton Communication. *Soliton Coding for Secured Optical Communication Link*, 17–39. USA, Springer.

Amiri, I.S., Alavi, S.E. and Idrus, S.M. (2015d) RF signal generation and wireless transmission using PANDA and Add/drop systems. *Journal of Computational and Theoretical Nanoscience (CTN)*.

Amiri, I.S., Alavi, S.E., Ahmad, H., Supa'at, A.S.M. and Fisal, N. (2014f) Numerical Computation of Solitonic Pulse Generation for Terabit/Sec Data Transmission. *Optical and Quantum Electronics*.

Amiri, I.S., Alavi, S.E., Bahadoran, M., Afroozeh, A. and Ahmad, H. (2015c) Nanometer Bandwidth Soliton Generation and Experimental Transmission within Nonlinear Fiber Optics Using an Add-Drop Filter System. *Journal of Computational and Theoretical Nanoscience (CTN)*, 12(2).

Amiri, I.S., Alavi, S.E., Fisal, N., Supa'at, A.S.M. and Ahmad, H. (2014b) All-Optical Generation of Two IEEE802.11n Signals for 2 × 2 MIMO-RoF via MRR System. *IEEE Photonics Journal*, 6(6).

Amiri, I.S., Alavi, S.E., Idrus, S.M. and Kouhnavard, M. (2014e) *Microring Resonator for Secured Optical Communication*. USA, Amazon.

Amiri, I.S., Alavi, S.E., Idrus, S.M., Afroozeh, A. and Ali, J. (2014c) *Soliton Generation by Ring Resonator for Optical Communication Application*. Hauppauge, NY 11788 USA, Nova Science Publishers.

Amiri, I.S., Alavi, S.E., Idrus, S.M., Nikoukar, A. and Ali, J. (2013d) IEEE 802.15.3c WPAN Standard Using Millimeter Optical Soliton Pulse Generated By a Panda Ring Resonator. *IEEE Photonics Journal*, 5(5): 7901912.

Amiri, I.S., Alavi, S.E., Idrus, S.M., Supa'at, A.S.M., Ali, J. and Yupapin, P.P. (2014g) W-Band OFDM Transmission for Radio-over-Fiber link Using Solitonic Millimeter Wave Generated by MRR. *IEEE Journal of Quantum Electronics*, 50(8): 622–628.

Amiri, I.S., Alavi, S.E., Soltanian, M.R.K. and Ahmad, H. (2015b) Tunable Channel Spacing of Soliton Comb Generation Using Add-drop Microring Resonators (MRRs). *Journal of Computational and Theoretical Nanoscience (CTN)*.

Amiri, I.S., Ali, J. and Yupapin, P.P. (2012c) Enhancement of FSR and Finesse Using Add/Drop Filter and PANDA Ring Resonator Systems. *International Journal of Modern Physics B*, 26(04): 1250034.

Amiri, I.S., Babakhani, S., Vahedi, G., Ali, J. and Yupapin, P. (2012d) Dark-Bright Solitons Conversion System for Secured and Long Distance Optical Communication. *IOSR Journal of Applied Physics (IOSR-JAP)*, 2(1): 43–48.

Amiri, I.S., Barati, B., Sanati, P., Hosseinnia, A., Mansouri Khosravi, H.R., Pourmehdi, S., Emami, A. and Ali, J. (2014c) Optical Stretcher of Biological Cells Using Sub-Nanometer Optical Tweezers Generated by an Add/Drop Microring Resonator System. *Nanoscience and Nanotechnology Letters*, 6(2): 111–117.

Amiri, I.S., Ebrahimi, M., Yazdavar, A.H., Gorbani, S., Alavi, S.E., Idrus, S.M. and Ali, J. (2014d) Transmission of data with orthogonal frequency division multiplexing technique for communication networks using GHz frequency band soliton carrier. *IET Communications*, 8(8): 1364–1373.

Amiri, I.S., Ghorbani, S. and Naraei, P. (2014h) Chaotic Carrier Signal Generation and Quantum Transmission Along Fiber Optics Communication Using Integrated Ring Resonators. *Quantum Matter*.

Amiri, I.S., Gifany, D. and Ali, J. (2013a) Entangled Photon Encoding Using Trapping of Picoseconds Soliton pulse. *IOSR Journal of Applied Physics (IOSR-JAP)*, 3(1): 25–31.

Amiri, I.S., Gifany, D. and Ali, J. (2013b) Long Distance Communication Using Localized Optical Soliton Via Entangled Photon. *IOSR Journal of Applied Physics (IOSR-JAP)*, 3(1): 32–39.

Amiri, I.S., Gifany, D. and Ali, J. (2013c) Ultra-short Multi Soliton Generation for Application in Long Distance Communication. *Journal of Basic and Applied Scientific Research (JBASR)*, 3(3): 442–451.

Amiri, I.S., Jalil, M.A., Mohamad, F.K., Ridha, N.J., Ali, J. and Yupapin, P.P. (2010) Storage of Optical Soliton Wavelengths Using NMRR. *International Conference on Experimental Mechanics (ICEM)*. Kuala Lumpur, Malaysia.

Amiri, I.S., Naraei, P. and Ali, J. (2014e) Review and Theory of Optical Soliton Generation Used to Improve the Security and High Capacity of MRR and NRR Passive Systems. *Journal of Computational and Theoretical Nanoscience (CTN)*, 11(9): 1875–1886.

Amiri, I.S., Nikmaram, M., Shahidinejad, A. and Ali, J. (2013) Generation of potential wells used for quantum codes transmission via a TDMA network communication system. *Security and Communication Networks*, 6(11): 1301–1309.

Amiri, I.S., Nikoukar, A. and Alavi, S.E. (2014a) *Soliton and Radio over Fiber (RoF) Applications*. Saarbrücken, Germany, LAP LAMBERT Academic Publishing.

Amiri, I.S., Nikoukar, A. and Ali, J. (2013a) GHz Frequency Band Soliton Generation Using Integrated Ring Resonator for WiMAX Optical Communication. *Optical and Quantum Electronics*, 46(9): 1165–1177.

Amiri, I.S., Nikoukar, A., Ali, J. and Yupapin, P.P. (2012e) Ultra-Short of Pico and Femtosecond Soliton Laser Pulse Using Microring Resonator for Cancer Cells Treatment. *Quantum Matter*, 1(2): 159–165.

Amiri, I.S., Nikoukar, A., Shahidinejad, A. and Anwar, T. (2014b) The Proposal of High Capacity GHz Soliton Carrier Signals Applied for Wireless Commutation. *Reviews in Theoretical Science*, 2(4): 320–333.

Amiri, I.S., Nikoukar, A., Shahidinejad, A., Anwar, T. and Ali, J. (2014a) Quantum Transmission of Optical Tweezers via Fiber Optic Using Half-Panda System. *Life Science Journal*, 10(12s): 391–400.

Amiri, I.S., Nikoukar, A., Shahidinejad, A., Ranjbar, M., Ali, J. and Yupapin, P.P. (2012f) Generation of Quantum Photon Information Using Extremely Narrow Optical Tweezers for Computer Network Communication. *GSTF Journal on Computing (joc)*, 2(1): 140.

Amiri, I.S., Rahim, F.J., Arif, A.S., Ghorbani, S., Naraei, P., Forsyth, D. and Ali, J. (2014b) Single Soliton Bandwidth Generation and Manipulation by Microring Resonator. *Life Science Journal*, 10(12s): 904–910.

Amiri, I.S., Raman, K., Afroozeh, A., Jalil, M.A., Nawi, I.N., Ali, J. and Yupapin, P.P. (2011d) Generation of DSA for security application. *Procedia Engineering*, 8: 360–365.

Amiri, I.S., Shahidinejad, A., Nikoukar, A., Ranjbar, M., Ali, J. and Yupapin, P.P. (2012g) Digital Binary Codes Transmission via TDMA Networks Communication System Using Dark and Bright Optical Soliton. *GSTF Journal on Computing (joc)*, 2(1): 12.

Amiri, I.S., Soltanian, M.R.K., Alavi, S.E. and Ahmad, H. (2015a) Multi Wavelength Mode-lock Soliton Generation Using Fiber Laser Loop Coupled to an Add-drop Ring Resonator. *Optical and Quantum Electronics*.

Amiri, I.S., Soltanmohammadi, S., Shahidinejad, A. and Ali, J. (2013e) Optical quantum transmitter with finesse of 30 at 800–nm central wavelength using microring resonators. *Optical and Quantum Electronics*, 45(10): 1095–1105.

Amiri, I.S., Vahedi, G., Shojaei, A., Nikoukar, A., Ali, J. and Yupapin, P.P. (2012h) Secured Transportation of Quantum Codes Using Integrated PANDA-Add/drop and TDMA Systems. *International Journal of Engineering Research and Technology (IJERT)*, 1(5).

Amiri, I.S., Zulkifli, M.Z. and Ahmad, H. (2014a) Soliton comb generation using add-drop ring resonators. *International Research Journal of Telecommunications and Information Technology*.

Arunvipas, P., Sangdao, C. and Phromloungsri, R. (2011) Spurious suppression and design based on microstrip open loop ring resonator bandpass filters. *IEICE Transactions on Electronics*, E94–C(9): 1447–1454.

Ayodeji, A.O., Amiri, I.S. and Fazeldehkordi, E. (2014) *A Machine Learning Approach to Phishing Detection and Defense*. USA, Elsevier.

Baehr-Jones, T., Hochberg, M., Wang, G., Lawson, R., Liao, Y., Sullivan, P., Dalton, L., Jen, A. and Scherer, A. (2005) Optical modulation and detection in slotted silicon waveguides. *Optics Express*, 13(14): 5216–5226.

Bahadoran, M., Ali, J. and Yupapin, P.P. (2013a) Graphical Approach for Nonlinear Optical Switching by PANDA Vernier Filter. *Photonics Technology Letters, IEEE*, 25(15): 1470–1473.

Bahadoran, M., Ali, J. and Yupapin, P.P. (2013b) Ultrafast all-optical switching using signal flow graph for PANDA resonator. *Applied Optics*, 52(12): 2866–2873.

Bahadoran, M., Amiri, I.S., Afroozeh, A., Ali, J. and Yupapin, P.P. (2011) Analytical Vernier Effect for Silicon Panda Ring Resonator. *National Science Postgraduate Conference, NSPC*, Universiti Teknologi Malaysia.

Bates, R.J. (2001) *Optical switching and networking handbook*. McGraw-Hill, Inc.

Beggs, D.M., Kampfrath, T., Rey, I., Krauss, T.F. and Kuipers, L. (2011) Controlling and switching slow light in photonic crystal waveguides. *Transparent Optical Networks (ICTON), 2011 13th International Conference on*, IEEE.

Bennink, R.S., Boyd, R.W., Stroud C.R., Jr. and Wong, V. (2001) Enhanced self-action effects by electromagnetically induced transparency in the two-level atom. *Physical Review A. Atomic, Molecular, and Optical Physics*, 63(3): 338041–338045.

Bigelow, M.S., Lepeshkin, N.N. and Boyd, R.W. (2003a) Observation of Ultraslow Light Propagation in a Ruby Crystal at Room Temperature. *Physical Review Letters*, 90(11): 113903.

Bigelow, M.S., Lepeshkin, N.N. and Boyd, R.W. (2003b) Superluminal and slow light propagation in a room-temperature solid. *Science*, 301(5630): 200.

Biswas, A. and Pati, G.S. (2011) Mathematical theory of slow light optical solitons. *Waves in Random and Complex Media*, 21(3): 456–468.

Blanchard, F., Sharma, G., Razzari, L., Ropagnol, X., Bandulet, H.C., Vidal, F., Morandotti, R., Kieffer, J.C., Ozaki, T., Tiedje, H., Haugen, H., Reid M. and Hegmann, F. (2011) Generation of intense terahertz radiation via optical methods. *IEEE Journal on Selected Topics in Quantum Electronics*, 17(1): 5–16.

Born, M. and Wolf, E. (1999) *Principles of optics: electromagnetic theory of propagation, interference and diffraction of light*. CUP Archive.

Born, M., Wolf, E. and Bhatia, A.B. (1999) *Principles of optics: electromagnetic theory of propagation, interference and diffraction of light*. Cambridge University Press.

Boyd, R.W. (1992) *Nonlinear Optics*. Academic Press, New York.

Boyd, R.W. and Gauthier, D.J. (2002) "Slow" and "fast" light. *Progress in Optics*, 43: 497–530.

Boyd, R.W., Gauthier, D.J. and Gaeta, A.L. (2006) Applications of slow light in telecommunications. *Optics and Photonics News*, 17(4): 19–23.

Brochu, P. and Pei, Q. (2010) Advances in dielectric elastomers for actuators and artificial muscles. *Macromolecular rapid communications*, 31(1): 10–36.

Burdea, G.C. and Langrana, N.A. (1995) Integrated virtual reality rehabilitation system, Google Patents.

Butcher, P.N. and Cotter, D. (1991) *The elements of nonlinear optics*. Cambridge University Press.

Capmany, J. and Muriel, M.A. (1990) A new transfer matrix formalism for the analysis of fiber ring resonators: compound coupled structures for FDMA demultiplexing. *Lightwave Technology, Journal of*, 8(12): 1904–1919.

Capmany, J., Ortega, B., Pastor, D. and Sales, S. (2005) Discrete-time optical processing of microwave signals. *Journal of Lightwave Technology*, 23(2): 702–723.

Carro-Lagoa, Á., Suárez-Casal, P., García-Naya, J.A., Fraga-Lamas, P., Castedo, L. and Morales-Méndez, A. (2013) Design and implementation of an OFDMA-TDD physical layer for WiMAX applications. *EURASIP Journal on Wireless Communications and Networking*, 2013(1): 1–19.

Chang, C.C. and Sirkis, J. (1996) Multiplexed optical fiber sensors using a single Fabry-Perot resonator for phase modulation. *Lightwave Technology, Journal of*, 14(7): 1653–1663.

Chang, K., Deb, S., Ganguly, A., Yu, X., Sah, S.P., Pande, P.P., Belzer, B. and Heo, D. (2012) Performance evaluation and design trade-offs for wireless network-on-chip architectures. *ACM Journal on Emerging Technologies in Computing Systems (JETC)*, 8(3): 23.

Chao, C.-Y., Ashkenazi, S., Huang, S.-W., O'Donnell, M. and Guo, L.J. (2007) High-frequency ultrasound sensors using polymer microring resonators. *Ultrasonics, Ferroelectrics and Frequency Control, IEEE Transactions on*, 54(5): 957–965.

Chen, D., Fetterman, H.R., Chen, A., Steier, W.H., Dalton, L.R., Wang, W. and Shi, Y. (1997) Demonstration of 110 GHz electro-optic polymer modulators. *Applied Physics Letters*, 70(25): 3335–3337.

Chiao, R.Y., Garmire, E. and Townes, C. (1964) Self-trapping of optical beams. *Physical Review Letters*, 13(15): 479.

Choi, J.M., Lee, R.K. and Yariv, A. (2002) Ring fiber resonators based on fused-fiber grating add-drop filters: application to resonator coupling. *Optics Letters*, 27(18): 1598–1600.

Chu, S. and Wong, S. (1982) Linear pulse propagation in an absorbing medium. *Physical Review Letters*, 48(11): 738–741.

Chuang, Y., Tseng, H. and Sheu, S. (2012) A Performance Study of Discrete-error-checking Scheme (DECS) with the Optimal Division Locations for IEEE 802.16–based Multi-hop Networks.

Cocorullo, G., Della Corte, F. and Rendina, I. (1999) Temperature dependence of the thermo-optic coefficient in crystalline silicon between room temperature and 550 K at the wavelength of 1523 nm. *Applied Physics Letters*, 74(22): 3338–3340.

Coexistence, I. (2009) Part 16: Air Interface for Fixed Broadband Wireless Access Systems.

Comtois, P. (2001) John Tyndall and the floating matter of the air. *Aerobiologia*, 17(3): 193–202.

Crutcher, S., Biswas, A., Aggarwal, M.D. and Edwards, M.E. (2005) Stationary temporal solitons in optical fiber and the swing effect of spatial solitons in two-dimensional devices. *Optics & Photonics 2005*, International Society for Optics and Photonics.

Cuomo, K.M. and Oppenheim, A.V. (1993) Chaotic signals and systems for communications. *Acoustics, Speech, and Signal Processing, 1993. ICASSP-93*, IEEE.

Daldosso, N. and Pavesi, L. (2009) Nanosilicon photonics. *Laser & Photonics Reviews*, 3(6): 508–534.

Desurvire, E., Bayart, D., Desthieux, B. and Bigo, S. (2002) *Erbium-doped fiber amplifiers*. John Wiley.

Dey, S. and Mandal, S. (2012) Enhancement of free spectral range in optical triple ring resonator: A vernier principle approach. *Recent Advances in Information Technology (RAIT), 2012 1st International Conference on*, 15–17 March 2012.

Dey, S.B., Mandal, S. and Jana, N. (2013) Quadruple optical ring resonator based filter on silicon-on-insulator. *Optik-International Journal for Light and Electron Optics*, 124(17): 2920–2927.

Dharmadhikari, A., Dharmadhikari, J. and Mathur, D. (2009) Visualization of focusing–refocusing cycles during filamentation in BaF 2. *Applied Physics B: Lasers and Optics*, 94(2): 259–263.

Diament, P. (1990) Wave transmission and fiber optics.

Diener, G. (1997) Energy transport in dispersive media and superluminal group velocities. *Physics Letters, Section A: General, Atomic and Solid State Physics*, 235(2): 118–124.

Djordjev, K., Choi, S.J. and Dapkus, P. (2002b) Vertically coupled InP microdisk switching devices with electroabsorptive active regions. *Photonics Technology Letters, IEEE*, 14(8): 1115–1117.

Djordjev, K., Choi, S.-J., Choi, S.-J. and Dapkus, P. (2002a) Microdisk tunable resonant filters and switches. *Photonics Technology Letters, IEEE*, 14(6): 828–830.

Dong, P., Liao, S., Feng, D., Liang, H., Zheng, D., Shafiiha, R., Kung, C.-C., Qian, W., Li, G. and Zheng, X. (2009) Low V$_{pp}$, ultralow-energy, compact, high-speed silicon electro-optic modulator. *Optics Express*, 17(25): 22484–22490.

El, G., Grimshaw, R. and Smyth, N. (2009) Transcritical shallow-water flow past topography: finite-amplitude theory. *Journal of Fluid Mechanics*, 640: 187–214.

Eliseev, P.G., Cao, H., Liu, C., Smolyakov, G.A. and Osiński, M. (2006) Nonlinear mode interaction as a mechanism to obtain slow/fast light in diode lasers.

Falcaro, P., Grosso, D., Amenitsch, H. and Innocenzi, P. (2004) Silica orthorhombic mesostructured films with low refractive index and high thermal stability. *The Journal of Physical Chemistry B*, 108(30): 10942–10948.

Fan, B.H., Zhang, Y.D. and Yuan, P. (2005) Observation of ultraslow light propagation in a ruby crystal at room temperature.

Fazeldehkordi, E., Amiri, I.S. and Akanbi, O.A. (2014) *Comparative Study of Multiple Black Hole Attacks Solution Methods in MANET Using AODV Routing Protocol*. Amazon.

Feng, M.Z., Sorin, W.V. and Tucker, R.S. (2009) Fast Light and the Speed of Information Transfer in the Presence of Detector Noise. *Photonics Journal, IEEE*, 1(3): 213–224.

Fischer, R., Neshev, D.N., Krolikowski, W., Kivshar, Y.S., Iturbe-Castillo, D., Chavez-Cerda, S., Meneghetti, R., Caetano, D.P. and Hickmann, J.M. (2006) Observation of spatial shift in interaction of dark nonlocal solitons, IEEE.

Fuji, T., Miyata, M., Kawato, S., Hattori, T. and Nakatsuka, H. (1997) Linear propagation of light investigated with a white-light Michelson interferometer. *JOSA B*, 14(5): 1074–1078.

Garrett, C.G.B. and McCumber, D.E. (1970) Propagation of a Gaussian light pulse through an anomalous dispersion medium. *Physical Review A*, 1(2): 305–313.

Gauthier, D.J. and Boyd, R.W. (2007) Fast light, slow light and optical precursors: What does it all mean?*Photonics Spectra*, 41(1): 82–84+86–88+90.

Gifany, D., Amiri, I.S., Ranjbar, M. and Ali, J. (2013) Logic Codes Generation and Transmission Using an Encoding-Decoding System. *International Journal of Advances in Engineering & Technology (IJAET)*, 5(2): 37–45.

Glaser, W. (1997) *Photonik für Ingenieure*. Verl. Technik.

Grover, R., Absil, P., Van, V., Hryniewicz, J., Little, B., King, O., Johnson, F., Calhoun, L. and Ho, P. (2001a) Vertically coupled GaAs-AlGaAs and GaInAsP-InP microring resonators. *Proceedings of Optical Fiber Communication Conference and Exhibit. Anaheim: IEEE.*

Grover, R., Absil, P.P., Van, V., Hryniewicz, J.V., Little, B.E., King, O., Calhoun, L.C., Johnson, F.G. and Ho, P.T. (2001b) Vertically coupled GaInAsP-InP microring resonators. *Optics Letters*, 26(8): 506–508.

Grover, R., Van, V., Ibrahim, T., Absil, P., Calhoun, L., Johnson, F., Hryniewicz, J. and Ho, P.T. (2002) Parallel-cascaded semiconductor microring resonators for high-order and wide-FSR filters. *Lightwave Technology, Journal of*, 20(5): 900–905.

Guarino, A., Poberaj, G., Rezzonico, D., Degl'Innocenti, R. and Günter, P. (2007) Electro–optically tunable microring resonators in lithium niobate. *Nature Photonics*, 1(7): 407–410.

Guidash, R., Lee, T.-H., Lee, P., Sackett, D., Drowley, C., Swenson, M., Arbaugh, L., Hollstein, R., Shapiro, F. and Domer, S. (1997) A 0.6/spl mu/m CMOS pinned photodiode color imager technology. *Electron Devices Meeting, 1997. IEDM'97. Technical Digest., International*, IEEE.

Hammond, B., Su, B., Mathews, J., Chen, E. and Schwartz, E. (2002) Integrated wavelength locker for tunable laser applications, IEEE.

Harke, A., Krause, M. and Mueller, J. (2005) Low-loss singlemode amorphous silicon waveguides. *Electronics Letters*, 41(25): 1377–1379.

Hasegawa, A. and Tappert, F. (1973a) Transmission of stationary nonlinear optical pulses in dispersive dielectric fibers. I. Anomalous dispersion. *Applied Physics Letters*, 23(3): 142–144.

Hasegawa, A. and Tappert, F. (1973b) Transmission of stationary nonlinear optical pulses in dispersive dielectric fibers. II. Normal dispersion. *Applied Physics Letters*, 23: 171.

Hecht, J. (1985) Victorian experiments and optical communications: Precursors of fiber-optic communications were invented a century ago, but no one at that time attempted to synthesize the fragmented knowledge. *Spectrum, IEEE*, 22(2): 69–73.

Hecht, J. (2004) *City of light: the story of fiber optics*. Oxford University Press, USA.

Hecht, J. (2010) Short history of laser development. *Optical Engineering* 49: 091002.

Heebner, J.E., Wong, V., Schweinsberg, A., Boyd, R.W. and Jackson, D.J. (2004) Optical transmission characteristics of fiber ring resonators. *Quantum Electronics, IEEE Journal of*, 40(6): 726–730.

Henker, R. (2010) *Investigation of the slow and fast light effect on the basis of SBS for application in optical communication and information system*. Doctor of Philosophy. Dublin Institute of Technology.

Henker, R., Schneider, T., Wiatrek, A., Lauterbach, K.U., Junker, M., Ammann, M.J. and Schwarzbacher, A.T. (2008a) Optimisation of optical signal delay in Slow-Light systems based on stimulated Brillouin scattering, IET.

Henker, R., Wiatrek, A., Lauterbach, K.U., Junker, M., Schneider, T., Ammann, M.J. and Schwarzbacher, A.T. (2008b) A review of slow-and fast-light based on stimulated brillouin scattering in future optical communication networks. *Komunikacie*, 10(4): 45–52.

Henry, C.H. (1982) Theory of the linewidth of semiconductor lasers. *Quantum Electronics, IEEE Journal of*, 18(2): 259–264.

Holman, R.L., Johnson, L.M.A. and Skinner, D.P. (1987) Desirability of electro-optic materials for guided-wave optics. *Optical Engineering*, 26(2): 262134–262134–.

Israwi, S. (2010) Variable depth KdV equations and generalizations to more nonlinear regimes. *ESAIM: Mathematical Modelling and Numerical Analysis*, 44(02): 347–370.

Junker, M., Schneider, T., Lauterbach, K.U., Henker, R., Ammann, M.J. and Schwarzbacher, A.T. (2007) High quality millimeter wave generation via stimulated Brillouin scattering, IEEE.

Katz, A. and Alfano, R. (1982) Pulse propagation in an absorbing medium. *Physical Review Letters*, 49(17): 1292.

Kawachi, M. (1990) Silica waveguides on silicon and their application to integrated-optic components. *Optical and Quantum Electronics*, 22(5): 391–416.

Keeler, G.A., Nelson, B.E., Agarwal, D., Debaes, C., Helman, N.C., Bhatnagar, A. and Miller, D.A. (2003) The benefits of ultrashort optical pulses in optically interconnected systems. *Selected Topics in Quantum Electronics, IEEE Journal of*, 9(2): 477–485.

Keiser, G. (2003) *Optical fiber communications*. Wiley Online Library.

Kempf, P. (2005) Enabling technology for analog integration. *System-on-Chip for Real-Time Applications, 2005. Proceedings. Fifth International Workshop on*, IEEE.

Kim, D.G., Choi, Y.W., Yi, J.C., Chung, Y., Ozturk, C. and Dagli, N. (2007) Multimode-interference-coupled ring resonators based on total-internal-reflection mirrors. *Japanese journal of applied physics. Pt. 1, Regular papers & short notes*, 46(1): 175–181.

Kim, H.D., Kang, S.-G. and Le, C.-H. (2000) A low-cost WDM source with an ASE injected Fabry-Perot semiconductor laser. *Photonics Technology Letters, IEEE*, 12(8): 1067–1069.

Knight, J.C., Broeng, J., Birks, T.A. and Russell, P.S.J. (1998) Photonic band gap guidance in optical fibers. *Science*, 282(5393): 1476–1478.

Kouhnavard, M., Afroozeh, A., Amiri, I.S., Jalil, M.A., Ali, J. and Yupapin, P.P. (2010a) New system of Chaotic Signal Generation Using MRR. *International Conference on Experimental Mechanics (ICEM)*. Kuala Lumpur, Malaysia.

Kouhnavard, M., Amiri, I.S., Jalil, M., Afroozeh, A., Ali, J. and Yupapin, P.P. (2010b) QKD via a quantum wavelength router using spatial soliton. *AIP Conference Proceedings*, 1347: 210–216.

Kronig, R.D. (1926) On the theory of dispersion of X-rays. *JOSA*, 12(6): 547–556.

Lange, C., Weis, E., Telekom, D. and Romero, S. (2012) An OFDMA-Based Optical Access Network Architecture Exhibiting Ultra-High Capacity and Wireline-Wireless Convergence. *IEEE Communications Magazine*: 3.

Li, Y. and Tong, L. (2008) Mach-Zehnder interferometers assembled with optical microfibers or nanofibers. *Optics Letters*, 33(4): 303–305.

Liang, D., Fiorentino, M., Bowers, J.E. and Beausoleil, R.G. (2011) Hybrid silicon ring lasers.

Lin, S. and Crozier, K.B. (2011) Planar silicon microrings as wavelength-multiplexed optical traps for storing and sensing particles. *Lab Chip*, 11(23): 4047–4051.

Lippmaa, E., Mägi, M., Samoson, A., Tarmak, M. and Engelhardt, G. (1981) Investigation of the structure of zeolites by solid-state high-resolution silicon-29 NMR spectroscopy. *Journal of the American Chemical Society*, 103(17): 4992–4996.

Little, B.E., Chu, S.T., Haus, H.A., Foresi, J. and Laine, J.-P. (1997a) Microring resonator channel dropping filters. *Lightwave Technology, Journal of*, 15(6): 998–1005.

Little, B.E., Chu, S.T., Haus, H.A., Foresi, J. and Laine, J.-P. (1997b) Microring resonator channel dropping filters. *Lightwave Technology, Journal of*, 15(6): 998–1005.

Liu, A., Jones, R., Liao, L., Samara-Rubio, D., Rubin, D., Cohen, O., Nicolaescu, R. and Paniccia, M. (2004) A high-speed silicon optical modulator based on a metal-oxide-semiconductor capacitor. *Nature*, 427(6975): 615–618.

Liu, A., Liao, L., Rubin, D., Nguyen, H., Ciftcioglu, B., Chetrit, Y., Izhaky, N. and Paniccia, M. (2007) High-speed optical modulation based on carrier depletion in a silicon waveguide. *Optics Express*, 15(2): 660–668.

Liu, K.-L. and Goan, H.-S. (2007) Non-Markovian entanglement dynamics of quantum continuous variable systems in thermal environments. *Physical Review A*, 76(2): 022312.

Liu, X., Osgood, R.M., Vlasov, Y.A. and Green, W.M. (2010) Mid-infrared optical parametric amplifier using silicon nanophotonic waveguides. *Nature Photonics*, 4(8): 557–560.

Longhi, S., Marano, M., Laporta, P., Svelto, O. and Belmonte, M. (2002) Propagation, manipulation, and control of picosecond optical pulses at 1.5 μm in fiber Bragg gratings. *Journal of the Optical Society of America B: Optical Physics*, 19(11): 2742–2757.

Lun, D., Zhang, B.-W., Mair, R.A., Zeng, K., Lin, J.Y., Jiang, H., Botchkarev, A., Kim, W., Morkoc, H. and Khan, M.A. (1998a) Optical properties and resonant modes in GaN/AlGaN and InGaN/GaN multiple quantum well microdisk cavities. *Photonics China'98*, International Society for Optics and Photonics.

Lun, D., Zhang, B.W., Mair, R.A., Zeng, K., Lin, J.Y., Jiang, H., Botchkarev, A., Kim, W., Morkoc, H. and Khan, M.A. (1998b) Optical properties and resonant modes in GaN/AlGaN and InGaN/GaN multiple quantum well microdisk cavities.

Madsen, C.K. and Zhao, J.H. (1999) *Optical filter design and analysis: a signal processing approach*. John Wiley & Sons, Inc.

Mandal, S., Dasgupta, K., Basak, T. and Ghosh, S. (2006) A generalized approach for modeling and analysis of ring-resonator performance as optical filter. *Optics Communications*, 264(1): 97–104.

Mårtensson, T., Svensson, C.P.T., Wacaser, B.A., Larsson, M.W., Seifert, W., Deppert, K., Gustafsson, A., Wallenberg, L.R. and Samuelson, L. (2004) Epitaxial III-V nanowires on silicon. *Nano Letters*, 4(10): 1987–1990.

McCall, S., Levi, A., Slusher, R., Pearton, S. and Logan, R. (1992) Whispering-gallery mode microdisk lasers. *Applied Physics Letters*, 60(3): 289–291.

McMillan, J.F., Yu, M., Kwong, D.L. and Wei Wong, C. (2010) Observation of four-wave mixing in slow-light silicon photonic crystal waveguides. *Optics Express*, 18(15): 15484–15497.

Melloni, A., Carniel, F., Costa, R. and Martinelli, M. (2001) Determination of bend mode characteristics in dielectric waveguides. *Journal of Lightwave Technology*, 19(4): 571.

Melnichuk, M. and Wood, L.T. (2010) Direct Kerr electro-optic effect in noncentrosymmetric materials. *Physical Review A*, 82(1): 013821.

Mirzaee, A. and Amiri, I.S. (2014). *Efficient Key Management for Symmetric Cryptography System*. USA, Amazon.

Mogilevtsev, D., Birks, T. and Russell, P.S.J. (1998) Group-velocity dispersion in photonic crystal fibers. *Optics Letters*, 23(21): 1662–1664.

Mok, J.T., De Sterke, C.M. and Eggleton, B.J. (2006) Delay-tunable gap-soliton-based slow-light system. *Optics Express*, 14(25): 11987–11996.

Mollenauer, L. and Smith, K. (1988) Demonstration of soliton transmission over more than 4000 kmin fiber with loss periodically compensated by Raman gain. *Optics Letters*, 13(8): 675–677.

Mollenauer, L.F., Stolen, R.H. and Gordon, J.P. (1980) Experimental observation of picosecond pulse narrowing and solitons in optical fibers. *Physical Review Letters*, 45(13): 1095–1098.

Mork, J.O.F., Xue, W., Chen Y., Blaaberg, S. and Sales, S. (2008) Slow and fast light effects in semiconductor waveguides for applications in microwave photonics *asia-pacific microwave photonics conference*. Gold Coast, Queens land Australia IEEE: 310–313.

Moslehi, B., Goodman, J.W., Tur, M. and Shaw, H.J. (1984) Fiber-optic lattice signal processing. *Proceedings of the IEEE*, 72(7): 909–930.

Nafea, H.B., Zaki, F.W. and Moustafa, H.E. (2013) Performance and Capacity Evaluation for Mobile WiMAX IEEE 802.16 m Standard. *Nature*, 1(1): 12–19.

Naraei, P., Amiri, I.S. and Saberi, I. (2014) *Optimizing IEEE 802.11i Resource and Security Essentials for Mobile and Stationary Devices*. Elsevier.

Narahara, K. and Nakagawa, S. (2010) Nonlinear traveling-wave field effect transistors for amplification of short electrical pulses. *IEICE Electronics Express*, 7(16): 1188–1194.

Narayanan, K. and Preble, S.F. (2010) Optical nonlinearities in hydrogenated-amorphous silicon waveguides. *Optics Express*, 18(9): 8998–9005.

Nasser, N., Fanjoux, G., Lantz, E. and Sylvestre, T. (2011) Tunable optical delay using parametric amplification in highly birefringent optical fibers. *Journal of the Optical Society of America B: Optical Physics*, 28(10): 2352–2357.

Neo, Y.S., Idrus, S.M., Rahmat, M.F., Alavi, S.E. and Amiri, I.S. (2014) Adaptive Control for Laser Transmitter Feedforward Linearization System. *IEEE Photonics Journal*, 6(4).

Nikoukar, A., Amiri, I.S. and Ali, J. (2013) Generation of Nanometer Optical Tweezers Used for Optical Communication Networks. *International Journal of Innovative Research in Computer and Communication Engineering*, 1(1): 77–85.

Nikoukar, A., Amiri, I.S., Alavi, S.E., Shahidinejad, A., Anwar, T., Supa'at, A.S.M., Idrus, S.M. and Teng, L.Y. (2014) Theoretical and Simulation Analysis of The Add/Drop Filter Ring Resonator Based on the Z-transform Method Theory. *The 2014 Third ICT International Student Project Conference (ICT-ISPC2014)*. Thailand, IEEE.

Nixon, R.H., Kemeny, S.E., Staller, C.O. and Fossum, E.R. (1995) 128 × 128 CMOS photodiode-type active pixel sensor with on-chip timing, control, and signal chain electronics. *IS&T/SPIE's Symposium on Electronic Imaging: Science & Technology*, International Society for Optics and Photonics.

Oda, K., Takato, N., Toba, H. and Nosu, K. (1988) A wide-band guided-wave periodic multi/demultiplexer with a ring resonator for optical FDM transmission systems. *Lightwave Technology, Journal of*, 6(6): 1016–1023.

Okamoto, K. (2006) *Fundamentals of optical waveguides*. Academic press.

Okawachi, Y., Bigelow, M.S., Sharping, J.E., Zhu, Z., Schweinsberg, A., Gauthier, D.J., Boyd, R.W. and Gaeta, A.L. (2005) Tunable all-optical delays via Brillouin slow light in an optical fiber. *Physical Review Letters*, 94(15): 153902.

Padmaraju, K. and Bergman, K. (2013) Resolving the thermal challenges for silicon microring resonator devices. *Lateral*, 60(1554.7): 1554.1558.

Padmaraju, K., Chan, J., Chen, L., Lipson, M. and Bergman, K. (2012a) Dynamic stabilization of a microring modulator under thermal perturbation. *Optical Fiber Communication Conference*, Optical Society of America.

Padmaraju, K., Chan, J., Chen, L., Lipson, M. and Bergman, K. (2012b) Thermal stabilization of a microring modulator using feedback control. *Optics Express*, 20(27): 27999–28008.

Padmaraju, K., Logan, D.F., Zhu, X., Ackert, J.J., Knights, A.P. and Bergman, K. (2013) Integrated thermal stabilization of a microring modulator. *Optics Express*, 21(12): 14342–14350.

Palais, J.C. (1988) *Fiber Optic Communications*. Prentice Hall.

Park, S.-J., Lee, C.-H., Jeong, K.-T., Park, H.-J., Ahn, J.-G. and Song, K.-H. (2004) Fiber-to-the-home services based on wavelength-division-multiplexing passive optical network. *Journal of Lightwave Technology*, 22(11): 2582.

Paul, K., Liu, Y.Z., Kellner, A.L., Williams, A.R., Lam, B.C. and Jiang, X. (1992) Design and fabrication of InGaAsP/InP waveguide modulators for microwave applications. *Aerospace Sensing*, International Society for Optics and Photonics.

Peng, K.-Q., Yan, Y.-J., Gao, S.-P. and Zhu, J. (2002) Synthesis of large-area silicon nanowire arrays via self-assembling nanoelectrochemistry. *Advanced Materials*, 14(16): 1164.

Pepeljugoski, P.K., Kash, J.A., Doany, F., Kuchta, D.M., Schares, L., Schow, C., Taubenblatt, M., Offrein, B.J. and Benner, A. (2010) Low power and high density optical interconnects for future supercomputers. *Optical Fiber Communication Conference*, Optical Society of America.

Perić, I., Blanquart, L., Comes, G., Denes, P., Einsweiler, K., Fischer, P., Mandelli, E. and Meddeler, G. (2006) The FEI3 readout chip for the ATLAS pixel detector. *Nuclear Instruments and Methods in Physics Research Section A: Accelerators, Spectrometers, Detectors and Associated Equipment*, 565(1): 178–187.

Poberaj, G., Koechlin, M., Sulser, F., Guarino, A., Hajfler, J. and Günter, P. (2009) Ion-sliced lithium niobate thin films for active photonic devices. *Optical Materials*, 31(7): 1054–1058.

Pornsuwancharoen, N., Sangwara, N. and Yupapin, P.P. (2010) Generalized fast and slow lights using multi-state microring resonators for optical wireless links. *Optik*, 121(19): 1721–1724.

Possley, N. and Upham, D.B. (2010) Recovery and transmission of return-to-zero formatted data using non-return-to-zero devices, Google Patents.

Provino, L., Dudley, J., Maillotte, H., Grossard, N., Windeler, R. and Eggleton, B. (2001) Compact broadband continuum source based on microchip laser pumped microstructured fibre. *Electronics Letters*, 37(9): 558–560.

Rabiei, P. and Gunter, P. (2004) Optical and electro-optical properties of submicrometer lithium niobate slab waveguides prepared by crystal ion slicing and wafer bonding. *Applied Physics Letters*, 85(20): 4603–4605.

Rabus, D.G., Bian, Z. and Shakouri, A. (2005) A GaInAsP-InP double-ring resonator coupled laser. *IEEE Photonics Technology Letters*, 17(9): 1770.

Razavi, B. (1996) A study of phase noise in CMOS oscillators. *Solid-State Circuits, IEEE Journal of*, 31(3): 331–343.

Reed, G.T., Mashanovich, G., Gardes, F. and Thomson, D. (2010) Silicon optical modulators. *Nature Photonics*, 4(8): 518–526.

Ridha, N.J., Mohamad, F.K., Amiri, I.S., Saktioto, Ali, J. and Yupapin, P.P. (2010a) Controlling Center Wavelength and Free Spectrum Range by MRR Radii. *International Conference on Experimental Mechanics (ICEM)*. Kuala Lumpur, Malaysia.

Ridha, N.J., Mohamad, F.K., Amiri, I.S., Saktioto, Ali, J. and Yupapin, P.P. (2010b) Soliton Signals and The Effect of Coupling Coefficient in MRR Systems. *International Conference on Experimental Mechanics (ICEM)*. Kuala Lumpur, Malaysia.

Roundy, S., Wright, P.K. and Rabaey, J. (2003) A study of low level vibrations as a power source for wireless sensor nodes. *Computer communications*, 26(11): 1131–1144.

Rybin, A. and Timonen, J. (2011) Nonlinear theory of slow light. *Philosophical Transactions of the Royal Society A: Mathematical, Physical and Engineering Sciences*, 369(1939): 1180–1214.

Saktioto, S., Ali, J., Hamdi, M. and Amiri, I.S. (2010a) Calculation and prediction of blood plasma glucose concentration. *ICAMN, International Conference*. Prince Hotel, Kuala Lumpur, Malaysia

Saktioto, S., Daud, S., Ali, J., Jalil, M.A., Amiri, I.S. and Yupapin, P.P. (2010b) FBG simulation and experimental temperature measurement.*ICEM*. Legend Hotel, Kuala Lumpur, Malaysia.

Saktioto, S., Hamdi, M., Amiri, I.S. and Ali, J. (2010c) Transition of diatomic molecular oscillator process in THz region. *International Conference on Experimental Mechanics (ICEM)*. Legend Hotel, Kuala Lumpur, Malaysia.

Saleh, B.E.A., Teich, M.C. and Saleh, B.E. (1991) *Fundamentals of photonics*. New York, John Wiley & Sons, Inc.

Sales, S., Xue, W., Mørk, J. and Gasulla, I. (2010) Slow and fast light effects and their applications to microwave photonics using semiconductor optical amplifiers. *IEEE Transactions on Microwave Theory and Techniques*, 58(11 PART 2): 3022–3038.

Sanati, P., Afroozeh, A., Amiri, I.S., Ali, J. and Chua, L.S. (2014) Femtosecond Pulse Generation using Microring Resonators for Eye Nano Surgery. *Nanoscience and Nanotechnology Letters*, 6(3): 221–226

Sander, J. and Hutter, K. (1991) On the development of the theory of the solitary wave. A historical essay. *Acta mechanica*, 86(1): 111–152.

Savage, N. (2002) Linking with light [high-speed optical interconnects]. *Spectrum, IEEE*, 39(8): 32–36.

Schweinsberg, A., Lepeshkin, N., Bigelow, M., Boyd, R. and Jarabo, S. (2006) Observation of superluminal and slow light propagation in erbium-doped optical fiber. *EPL (Europhysics Letters)*, 73(2): 218.

Schwelb, O. (1998) Generalized analysis for a class of linear interferometric networks. I. Analysis. *Microwave Theory and Techniques, IEEE Transactions on*, 46(10): 1399–1408.

Scolari, L., Alkeskjold, T., Riishede, J., Bjarklev, A., Hermann, D., Anawati, A., Nielsen, M. and Bassi, P. (2005) Continuously tunable devices based on electrical control of dual-frequency liquid crystal filled photonic bandgap fibers. *Optics Express*, 13(19): 7483–7496.

Shahidinejad, A., Amiri, I.S. and Anwar, T. (2014) Enhancement of Indoor Wavelength Division Multiplexing-Based Optical Wireless Communication Using Microring Resonator. *Reviews in Theoretical Science*, 2(3): 201–210.

Shahidinejad, A., Soltanmohammadi, S., Amiri, I.S. and Anwar, T. (2014) Solitonic Pulse Generation for Inter-Satellite Optical Wireless Communication. *Quantum Matter*, 3(2): 150–154.

Shen, Y.-R. (1984) The principles of nonlinear optics. *New York, Wiley-Interscience, 1984*, 575, p. 1.

Sherman, G.C. and Oughstun, K.E. (1981) Description of pulse dynamics in Lorentz media in terms of the energy velocity and attenuation of time-harmonic waves. *Physical Review Letters*, 47(20): 1451–1454.

Shojaei, A.A. and Amiri, I.S. (2011a) DSA for Secured Optical Communication. *International Conference for Nanomaterials Synthesis and Characterization (INSC)*. Kuala Lumpur, Malaysia.

Shojaei, A.A. and Amiri, I.S. (2011b) Soliton for Radio wave generation. *International Conference for Nanomaterials Synthesis and Characterization (INSC)*. Kuala Lumpur, Malaysia.

Sirawattananon, C., Bahadoran, M., Ali, J., Mitatha, S. and Yupapin, P.P. (2012) Analytical Vernier Effects of a PANDA Ring Resonator for Microforce Sensing Application. *Nanotechnology, IEEE Transactions on*, 11(4): 707–712.

Snyder, A.W. and Love, J.D. (1983) *Optical Waveguide Theory*. Springer.

Soares, J., Beliaev, D., Enderlein, R., Scolfaro, L., Saito, M. and Leite, J. (1995) Photoreflectance investigations of semiconductor device structures. *Materials Science and Engineering: B*, 35(1): 267–272.

So-In, C., Jain, R. and Tamimi, A.-K. (2009) Scheduling in IEEE 802.16 e mobile WiMAX networks: key issues and a survey. *Selected Areas in Communications, IEEE Journal on*, 27(2): 156–171.

Soltani, M., Yegnanarayanan, S. and Adibi, A. (2007) Ultra-high Q planar silicon microdisk resonators for chip-scale silicon photonics. *Optics Express*, 15(8): 4694–4704.

Soltanian, S.M.R.K. and Amiri, I.S. (2014) *Detection and Defeating Distributed Denial of Service (DDoS) Attacks*. USA, Amazon.

Song, K.Y., Herráez, M.G. and Thévenaz, L. (2005) Observation of pulse delaying and advancement in optical fibers using stimulated Brillouin scattering. *Optics Express*, 13(1): 82–88.

Soref, R. (2006) The past, present, and future of silicon photonics. *Selected Topics in Quantum Electronics, IEEE Journal of*, 12(6): 1678–1687.

Spyropoulou, M., Pleros, N. and Miliou, A. (2011) SOA-MZI-based nonlinear optical signal processing: A frequency domain transfer function for wavelength conversion, clock recovery, and packet envelope detection. *Quantum Electronics, IEEE Journal of*, 47(1): 40–49.

Srikanth, S., Pandian, M. and Fernando, X. (2012) Orthogonal frequency division multiple access in WiMAX and LTE: a comparison. *Communications Magazine, IEEE*, 50(9): 153–161.

Stanton, T. and Ostrovsky, L. (1998) Observations of highly nonlinear internal solitons over the continental shelf. *Geophysical Research Letters*, 25(14): 2695–2698.

Stegeman, G.I. and Segev, M. (1999) Optical spatial solitons and their interactions: universality and diversity. *Science*, 286(5444): 1518–1523.

Stenner, M.D., Gauthier, D.J. and Neifeld, M.A. (2003) The speed of information in a 'fast-light' optical medium. *Nature*, 425(6959): 695–698.

Stolen, R. and Lin, C. (1978) Self-phase-modulation in silica optical fibers. *Physical Review A*, 17(4): 1448.

Stratmann, M. and Mitschke, F. (2005) Chains of temporal dark solitons in dispersion-managed fiber. *Physical review. E, Statistical, nonlinear, and soft matter physics*, 72(6 Pt 2): 066616.

Su, Y., Liu, F., Li, Q., Zhang, Z. and Qiu, M. (2007) System performance of slow-light buffering and storage in silicon nano-waveguide. *Asia Pacific Optical Communications*, International Society for Optics and Photonics.

Suchat, S., Pornsuwancharoen, N. and Yupapin, P. (2010) Continuous variable quantum key distribution via a simultaneous optical-wireless up-down-link system. *Optik-International Journal for Light and Electron Optics*, 121(17): 1540–1544.

Suhailin, F., Ali, J., Yupapin, P.P., Fujii, Y., Ahmad, H. and Harun, S.W. (2009) Stopping and storing light pulses within a fiber optic ring resonator. *Chinese Optics Letters*, 7(9): 778–780.

Suwanpayak, N., Songmuang, S., Jalil, M.A., Amiri, I.S., Naim, I., Ali, J. and Yupapin, P.P. (2010) Tunable and storage potential wells using microring resonator system for bio-cell trapping and delivery. *AIP Conference Proceedings*, 1341: 289–291.

Tadigadapa, S. and Mateti, K. (2009) Piezoelectric MEMS sensors: state-of-the-art and perspectives. *Measurement Science and Technology*, 20(9): 092001.

Takagi, H., Maeda, R., Hosoda, N. and Suga, T. (1999) Room-temperature bonding of lithium niobate and silicon wafers by argon-beam surface activation. *Applied Physics Letters*, 74(16): 2387–2389.

Tang, X.Y. and Shukla, P.K. (2007) Solution of the one-dimensional spatially inhomogeneous cubic-quintic nonlinear Schrödinger equation with an external potential. *Physical Review A – Atomic, Molecular, and Optical Physics*, 76(1).

Thammawongsa, N., Moongfangklang, N., Mitatha, S. and Yupapin, P.P. (2012) Novel Nano-Antenna System Design Using Photonic Spin in a Panda Ring Resonator. *Progress In Electromagnetics Research*, 31: 75–87.

Thévenaz, L. (2008) Slow and fast light in optical fibres. *Nature Photonics*, 2(8): 474–481.

Thévenaz, L., Song, K.-Y., Chin, S.-H. and Gonzalez-Herraez, M. (2007) Light controlling light in an optical fibre: from very slow to faster-than-light speed. *Intelligent Signal Processing, 2007. WISP 2007. IEEE International Symposium on*, IEEE.

Toll, J.S. (1956) Causality and the dispersion relation: logical foundations. *Physical Review*, 104(6): 1760.

Tovar, A.A. and Casperson, L.W. (1995) Gaussian beam optical systems with high gain or high loss media. *Microwave Theory and Techniques, IEEE Transactions on*, 43(8): 1857–1862.

Tsolkas, D., Xenakis, D., Passas, N. and Merakos, L. (2012) Next Generation Cognitive Cellular Networks, LTE, WiMAX and Wireless Broadband Access. *Cognitive Radio and its Application for Next Generation Cellular and Wireless Networks*, 307–330, Springer.

Tucker, R.S., Ku, P.-C. and Chang-Hasnain, C.J. (2005) Slow-light optical buffers: capabilities and fundamental limitations. *Journal of Lightwave Technology*, 23(12): 4046.

Tunsiri, S., Kanthavong, S., Mitatha, S. and Yupapin, P. (2012) Optical-Quantum Security using Dark-Bright Soliton Conversion in a Ring Resonator System. *Procedia Engineering*, 32: 475–481.

Uranus, H.P., Zhuang, L., Roeloffzen, C.G.H. and Hoekstra, H.J.W.M. (2007) Pulse advancement and delay in an integrated-optical two-port ring-resonator circuit: Direct experimental observations. *Optics Letters*, 32(17): 2620–2622.

Wang, T., Liu, F., Wang, J., Tian, Y., Zhang, Z., Ye, T., Qiu, M. and Su, Y. (2009a) Pulse delay and advancement in SOI microring resonators with mutual mode coupling. *Lightwave Technology, Journal of*, 27(21): 4734–4743.

Wang, X., Tian, H., Li, C. and Ji, Y. (2009b) Tunable slow light by electro-eptic effect in polymer photonic crystal waveguide. *Guangxue Xuebao/Acta Optica Sinica*, 29(5): 1374–1378.

Wang, Z., Kravtsov, K.S., Huang, Y.-K. and Prucnal, P.R. (2011) Optical FFT/IFFT circuit realization using arrayed waveguide gratings and the applications in all-optical OFDM system. *Optics Express*, 19(5): 4501–4512.

Wiatrek, A., Henker, R., Preußler, S. and Schneider, T. (2009) 1.4 Bit Delay and Pulse Compression Based on Brillouin Optical Signal Processing, Optical Society of America.

Wise, F.W. (2001) Spatiotemporal solitons in quadratic nonlinear media. *Pramana*, 57(5): 1129–1138.

Xia, F., Sekaric, L. and Vlasov, Y. (2006) Ultracompact optical buffers on a silicon chip. *Nature Photonics*, 1(1): 65–71.

Xu, Q., Fattal, D. and Beausoleil, R.G. (2008) Silicon microring resonators with 1.5-μm radius. *Optics Express*, 16(6): 4309–4315.

Xu, Q., Schmidt, B., Pradhan, S. and Lipson, M. (2005) Micrometre-scale silicon electro-optic modulator. *Nature*, 435(7040): 325–327.

Yan, S.L. (2010) Enhancement of chaotic carrier bandwidth in a semiconductor laser transmitter using self-phase modulation in an optical fiber external round cavity. *Chinese Science Bulletin*, 55(11): 1007–1012.

Yariv, A. (2000) Universal relations for coupling of optical power between microresonators and dielectric waveguides. *Electronics Letters*, 36(4): 321–322.

Zadok, A., Eyal, A. and Tur, M. (2011) Stimulated Brillouin scattering slow light in optical fibers. *Applied Optics*, 50(25): 11.

Zalevsky, Z., Shemer, A., Eckhouse, V., Mendlovic, D. and Zach, S. (2005) Radio frequency photonic filter for highly resolved and ultrafast information extraction. *Journal of the Optical Society of America A: Optics and Image Science, and Vision*, 22(8): 1668–1677.

Zeinalinezhad, A., Pourmand, S.E., Amiri, I.S. and Afroozeh, A. (2014) Stop Light Generation using Nano Ring Resonators for ROM. *Journal of Computational and Theoretical Nanoscience (CTN)*.

Zhang, X.F., He, W.Q. and Zhang, P. (2011) Controllable Optical Solitons in Optical Fiber System with Distributed Coefficients. *Communications in Theoretical Physics*, 55: 681.

Appendices

Optical transfer function of SRR using Z-transform method

The fraction of light passing via the throughput: $\quad C = \sqrt{(1-\gamma)(1-k)}$

The fraction of light passing via the crossed route: $\quad S = \sqrt{(1-\gamma)k}$

Z-transform parameter: $\quad z^{-1} = \exp(-iKn_gL)$

Roundtrip loss: $\quad x = \exp(-\alpha L/2)$

Input node: $\quad E_1 = E_{in}$

Through node: $\quad E_3 = E_{out}$

Through port: $\quad H_{31} = E_3(z)/E_1(z)$

Gain from input port toward through port:

$$L(SRR) = \sqrt{(1-\gamma)(1-k)}\,e^{-\frac{\alpha L}{2}-iKnL} = C\xi$$

Transmission coefficient: $\quad \xi = x\,z^{-1}$

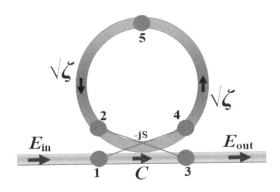

The single ring resonator

Roundtrip loss coefficient: $x = e^{-\alpha L/2}$

Z-transform parameter: $z^{-1} = e^{-i\varphi}$

Through paths: T_i

The symbol Δ denotes the Signal Fellow Graph determinant:

$$\Delta = 1 - \sum_i L_i + \sum_{i \neq j} L_i \cdot L_j - \sum_{i \neq j \neq k} L_i \cdot L_j \cdot L_k + \cdots$$

Forward path transmittance $(1 \to 3)$: $T_{1,\text{out}}(SRR) = C$

Delta determinant for through port: $\Delta_{1,\text{out}}(SRR) = 1 - L = 1 - C\xi$

Forward path transmittance $(1 \to 4\text{-}5\text{-}2 \to 3)$: $T_2(SRR) = -S^2\xi$

Delta determinant for through port path: $\Delta_{2,\text{out}}(SRR) = 1$

Transfer function or input-output transmittance relationship: $H = \frac{1}{\Delta} \sum_{i=1}^{n} T_i \Delta_i$

Transfer function for through port:

$$H_{31}(SRR) = \frac{E_{\text{out}}}{E_{\text{in}}} = \frac{C - \xi}{1 - C\xi} = \frac{\sqrt{(1-\gamma)(1-k)} - e^{-\frac{\alpha L}{2} - iKnL}}{1 - \sqrt{(1-\gamma)(1-k)} - e^{-\frac{\alpha L}{2} - iKnL}}$$

The normalized intensity: $I = H \cdot H^*$

Normalized intensity relation for output/input port of SRR:

$$I(SRR) = \frac{C^2 + x^2 - 2C\xi \cos\varphi}{1 - C^2 x^2 - 2C\xi \cos\varphi}$$

Optical transfer function of add-drop filter using Z-transform method

Coupling factor: k_i

Portion of light pulse crossing via the direct path: $C_i = \sqrt{(1-\gamma)(1-k_i)}$

Portion of light running on the crossing path: $S_i = \sqrt{(1-\gamma)\,k_i}$

Loss coefficient in each round trip: $x = \exp(-\alpha\pi R)$

Z-transform parameter: $z^{-1} = \exp(-i2\pi n_g L/\lambda)$

Light pulse touches the $2 \to 4 \to 5 \to 7$ photonics nodes as: $L(ADF) = C_1 C_2 \xi_1$

The forward pathway transmittance from: $(1\to4\to5\to8)$: $T_{1,\mathrm{drop}}(ADF) = -S_1 S_2 \sqrt{\xi}$

Delta determinant for this trail is: $\Delta_{1,\mathrm{drop}}(ADF) = 1$

The symbol Δ: $\Delta = 1 - \sum_i L_i + \sum_{i \neq j} L_i \cdot L_j - \sum_{i \neq j \neq k} L_i \cdot L_j \cdot L_k + \cdots$

Transfer function or input-output transmittance relationship: $H = \dfrac{1}{\Delta} \sum_{i=1}^{n} T_i \Delta_i$

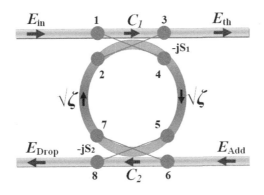

Signal flow graph of an add/drop filter system

Transfer function for drop port:

$$H_{\text{drop}}(ADF) = \frac{E_8}{E_1} = \frac{-S_1 S_2 \sqrt{\xi}}{1 - C_1 C_2 \xi}$$

$$= \frac{-\sqrt{(1-\gamma)k_1}\sqrt{(1-\gamma)k_2}\,e^{(-\alpha L/4)}e^{(-Kn_g L/2)}}{1 - \sqrt{(1-\gamma)(1-k_1)}\sqrt{(1-\gamma)(1-k_2)}\,e^{(-\alpha L/2)}e^{(-Kn_g L)}}$$

The intensity functions for drop port of add-drop filter:

$$I_{\text{drop}}(ADF) = H_{\text{drop}} \times H_{\text{drop}}^* = \left|\frac{E_8}{E_1}\right|^2 = \frac{S_1^2 S_2^2 x^2}{(1 - C_1 C_2 x)^2 + 4C_1 C_2 x \sin^2\left(\frac{\varphi}{2}\right)}$$

A direct route from node 1 to 3: $T_{1,th}(ADF) = C_1$

Delta determinant function for direct path $\Delta_{1,th}(ADF) = 1 - L = 1 - C_1 C_2 \xi$

The track passes through port waveguide $(1 \to 4 \to 5 \to 7 \to 2 \to 3)$ nodes:

$$T_{2,th}(ADF) = -S_1^2 C_2 \zeta$$

The delta determinant: $\Delta_{2,th}(ADF) = 1$

Optical transfer functions for through port of add/drop filter:

$$H_{th}(ADF) = \frac{E_3}{E_1} = \frac{C_1 - C_2 \xi}{1 - C_1 C_2 \xi}$$

$$= \frac{\sqrt{(1-\gamma)(1-k_1)} - \sqrt{(1-\gamma)(1-k_2)}\,e^{(-\alpha L/2)}e^{(-Kn_g L)}}{1 - \sqrt{(1-\gamma)(1-k_1)}\sqrt{(1-\gamma)(1-k_2)}\,e^{(-\alpha L/2)}e^{(-Kn_g L)}}$$

The intensity functions for throughput of add-drop filter:

$$I_{th}(ADF) = H_{th} \times H_{th}^* = \left|\frac{E_3}{E_1}\right|^2 = 1 - \frac{S_1^2(1 - C_2^2 x^2)}{(1 - C_1 C_2 x)^2 + 4C_1 C_2 x \sin^2(\frac{\varphi}{2})}$$

Optical transfer function of PANDA ring resonator using Z-transform method

Fraction of light passed through the throughput path: $C_i = \sqrt{(1 - \gamma_i)(1 - k_i)}$

Fraction path via cross path: $S_i = \sqrt{(1 - \gamma_i)k_i}$

Input node: $E_1(z)$

Through node: $E_3(z)$

Optical transfer function for through port of PANDA: $H_{31}^1 = E_3(z)/E_1(z)$

Loop gains of the PANDA from input port toward through port:

$$L_1 = C_1 C_2 C_R C_L \xi^N \qquad\qquad L_2 = L_R = C_R \xi_R^{NR}$$

$$L_3 = L_L = C_L \xi_L^{NL} \qquad\qquad L_4 = L_{1R} = -C_1 C_2 C_L S_R^2 \xi_R^{NR} \xi^N$$

$$L_5 = L_{1L} = -C_1 C_2 C_R S_L^2 \xi_L^{NL} \xi^N \quad L_6 = C_1 C_2 S_R^2 S_L^2 \xi_L^{NL} \xi_R^{NR} \xi^N$$

Roundtrip loss: $X_q = \exp(-\alpha L_1/2)$

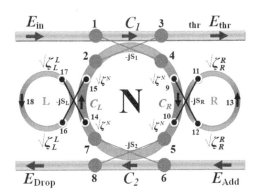

Z-transform parameter: $Z^{-p} = \exp(-j\phi p)$

Phase shift: $\phi = k n_{\text{eff}} L$

Multiplication of one roundtrip loss into Z-transform parameter coefficient:

$$\xi_q^p \equiv X_q Z^{-p}$$

Integer resonant numbers of each ring in a PANDA: p

Transmittances of two non-touching loops:

$$L_{12} = L_1 \cdot L_R = C_1 C_2 C_R^2 C_L \xi_R^{NR} \xi^N,$$

$$L_{13} = L_1 \cdot L_L = C_1 C_2 C_L^2 C_R \xi_L^{NL} \xi^N,$$

$$L_{23} = L_R \cdot L_L = C_L C_R \xi_L^{NL} \xi_R^{NR},$$

$$L_{25} = L_R \cdot L_5 = -C_1 C_2 C_R^2 S_L^2 \xi_L^{NL} \xi_R^{NR} \xi^N,$$

$$L_{34} = L_L \cdot L_4 = -C_1 C_2 C_L^2 S_R^2 \xi_L^{NL} \xi_R^{NR} \xi^N,$$

For three non-touching loops, one possible product of transmittances can be found as:

$$L_{123} = L_1 \cdot L_2 \cdot L_3 = C_1 C_2 C_L^2 C_R^2 \xi_L^{NL} \xi_R^{NR} \xi^N,$$

The forward path transmittance for direct path $1 \rightarrow 3$: $T_{1,thr}(PANDA) = C_1$

Non-touching delta for this track is:

$$\Delta_{1,thr}(PANDA) = 1 - \sum_{n=1}^{6} L_n + L_{12} + L_{13} + L_{23} + L_{25} + L_{34} - L_{123}$$

The forward path transmittance for $1 \rightarrow 4\text{-}9\text{-}10\text{-}5\text{-}7\text{-}14\text{-}15\text{-}2 \rightarrow 3$:

$$T_{2,thr}(PANDA) = -C_2 C_R C_L S_{15}^2 \xi^N$$

Non-touching (right and left small rings) delta for this track:

$$\Delta_{2,thr}(PANDA) = 1 - L_2 - L_3 + L_{23}$$

The forward path transmittance for $1 \rightarrow 4\text{-}9\text{-}12\text{-}13\text{-}11\text{-}10\text{-}5\text{-}7\text{-}14\text{-}15\text{-}2 \rightarrow 3$ nodes:

$$T_{3,thr}(PANDA) = C_2 C_L S_1^2 S_R^2 \xi^N \xi_R^{NR}$$

If the left small ring does not touch this route the path determinant is:

$$\Delta_{3,thr}(PANDA) = 1 - L_3$$

The forward path transmittance for $1 \rightarrow$ 4-9-12-13-11-10-5-7-14-17-18-16-15-2 $\rightarrow$ 3 nodes:

$$T_{4,thr}(PANDA) = -C_2 S_1^2 S_R^2 S_L^2 \xi^N \xi_R^{NR} \xi_L^{NL}$$

The delta determinant for this path: $\Delta_{4,thr}(PANDA) = 1$

The forward path transmittance for $1 \rightarrow$ 4-9-10-5-7-14-17-18-16-15-2 $\rightarrow$ 3 nodes:

$$T_{5,thr}(PANDA) = C_2 C_R S_1^2 S_L^2 \xi^N \xi_L^{NL}$$

Determinant's delta: $\Delta_{5,thr}(PANDA) = 1 - L_2$

Transfer function of the PANDA at the through port:

$$H_{thr}(PANDA) = \frac{E_3}{E_1} = \frac{\begin{aligned}\{C_1(1 - C_R\xi_R^{NR} - C_L\xi_L^{NL} + C_R C_L\xi_R^{NR}\xi_L^{NL}) \\ + C_2\xi^N \left(C_R\xi_L^{NL} + C_L\xi_R^{NR} - C_L C_R - \xi_L^{NL}\xi_R^{NR}\right)\}\end{aligned}}{\begin{aligned}\{1 - C_1 C_2 C_R C_L\xi^N - C_R\xi_R^{NR} - C_L\xi_L^{L} \\ + C_1 C_2 C_L S_R^2 \xi_R^{NR} \xi^N + C_1 C_2 C_R S_L^2 \xi_L^{NL} \xi^N \\ - C_1 C_2 S_R^2 S_L^2 \xi_L^{NL} \xi_R^{NR} \xi^N + C_1 C_2 C_L^2 C_L \xi_R^{NR} \xi^N \\ + C_1 C_2 C_L^2 C_R \xi_L^{NL} \xi^N + C_L C_R \xi_L^{NL} \xi_R^{NR} - C_1 C_2 \\ \times C_R^2 S_L^2 \xi_L^{NL} \xi_R^{NR} \xi^N - C_1 C_2 C_L^2 S_R^2 \xi_L^{NL} \xi_R^{NR} \xi^N \\ + C_1 C_2 C_L^2 C_R^2 \xi_L^{NL} \xi_R^{NR} \xi^N\}\end{aligned}}$$

DROP PORT'S OPTICAL TRANSFER FUNCTION

Optical transfer function for drop port of PANDA: $H_{81}^1 = E_8(z)/E_1(z)$

Input node: $E_1(z)$

Drop node: $E_8(z)$

The forward path transmittance for paths 1-4-9-10-5 $\rightarrow$ 8:

$$T_{1,drp}(PANDA) = -C_R S_1 S_2 \sqrt{\xi^N}$$

$$\Delta_{1,drp}(PANDA) = 1 - L_2 - L_3 + L_{23}$$

The forward path transmittance for path 1-4-9-12-13-11-10-5 $\rightarrow$ 8:

$$T_{2,drp}(PANDA) = S_R^2 S_1 S_2 \xi_R^{NR} \sqrt{\xi^N}$$

$$\Delta_{2,drp}(PANDA) = 1 - L_3$$

Optical transfer function of the PANDA at the drop port:

$$
\begin{aligned}
H_{\mathrm{drp}}(PANDA) = \frac{E_8}{E_1}L = \frac{\{S_1 S_2 \sqrt{\xi^N}(C_R C_L \xi_L^{NL} - C_L \xi_L^{NL}\xi_R^{NR} + \xi_R^{NR} - C_R)\}}{\{1 - C_1 C_2 C_R C_L \xi^N - C_R \xi_R^{NR} - C_L \xi_L^L} \\
+ C_1 C_2 C_L S_R^2 \xi_R^{NR}\xi^N + C_1 C_2 C_R S_L^2 \xi_L^{NL}\xi^N \\
- C_1 C_2 S_R^2 S_L^2 \xi_L^{NL}\xi_R^{NR}\xi^N + C_1 C_2 C_R^2 C_L \xi_R^{NR}\xi^N \\
+ C_1 C_2 C_L^2 C_R \xi_L^{NL}\xi^N + C_L C_R \xi_L^{NL}\xi_R^{NR} - C_1 C_2 \\
\times C_R^2 S_L^2 \xi_L^{NL}\xi_R^{NR}\xi^N - C_1 C_2 C_L^2 S_R^2 \xi_L^{NL}\xi_R^{NR}\xi^N \\
+ C_1 C_2 C_L^2 C_R^2 \xi_L^{NL}\xi_R^{NR}\xi^N\}
\end{aligned}
$$

******

The MATLAB programs are included in Appendix D to K, which can be found on the book webpage on the CRC Press website: www.crcpress.com/9781138027831.

******